Electrochemical Biosensors

Electrochemical Biosensors

Applications in Diagnostics, Therapeutics, Environment, and Food Management

Pranveer Singh

ELSEVIER

ACADEMIC PRESS

An imprint of Elsevier

Academic Press is an imprint of Elsevier
125 London Wall, London EC2Y 5AS, United Kingdom
525 B Street, Suite 1650, San Diego, CA 92101, United States
50 Hampshire Street, 5th Floor, Cambridge, MA 02139, United States
The Boulevard, Langford Lane, Kidlington, Oxford OX5 1GB, United Kingdom

Library of Congress Cataloging-in-Publication Data
A catalog record for this book is available from the Library of Congress

British Library Cataloguing-in-Publication Data
A catalogue record for this book is available from the British Library

ISBN 978-0-323-90632-6

For information on all Academic Press publications
visit our website at https://www.elsevier.com/books-and-journals

Publisher: Billie Jean Fernandez
Acquisitions Editor: Glyn Jones
Editorial Project Manager: Billie Jean Fernandez
Production Project Manager: Glyn Jones
Cover Designer: Matthew Limbert

Typeset by STRAIVE, India

Contents

Preface

The ever-increasing demands for food, water, energy, health, and infrastructure require that sensor-based solutions to these are both sustainable and reproducible. Recently, electrochemical sensors have shown impressive progress in the field of biosensing and analytical chemistry. Highly selective electrochemical sensors capable of picomolar sensitivity and high-throughput analysis with low sample requirements have already been demonstrated. Improved electrochemical sensors incorporating nanomaterials are on the horizon. Nanomaterials with useful attributes such as good optical, catalytic, magnetic, and antimicrobial properties not only enhance sensing but also perform rapid capture and detection of analytes.

The current book titled "Electrochemical Nano-biosensors: Applications in Diagnostics, Therapeutics, Environment, and Food management" is divided into five chapters. Chapter 1 focusses on the details of different types of sensors with emphasis on electrochemical biosensors. Different versions of electrochemical biosensors are described based on the electrochemical techniques involved. Chapter 2 describes the role of different nanomaterials and their hybrid play in the structure, synthesis, and function of an electrochemical biosensor. Chapter 3 discusses the application of electrochemical biosensors in the biomonitoring of clinical biomarkers related to cancers, viral and bacterial pathogens, metabolic disorders, genetic abnormalities, and neurological disorders. Chapter 4 narrates the biomonitoring of various environmental pollutants and contaminants, including organic and inorganic contaminants, heavy metals, water-borne pathogens, and agricultural pollutants such as pesticides, fungicides, soil microbes, and pathogens. It also details the use of nano-based agricultural products such as nanopesticides and nanofertilizers. Chapter 5 discusses in entirety the biomonitoring of food adulterants, toxicants, allergens, pathogens, and spurious additives. Additionally, all chapters delineate the existing state of relevant techniques and protocols and future directions for their improvisations. Literatures spanning more than 2 decades are vetted to elucidate the topics covered in the chapters. This book critically analyzes the issues and challenges related to the application of nanomaterial-incorporated electrochemical biosensors in real-world settings, namely, universal applicability in resource-limited sets-ups in remote areas, ease of integration with other sensing platforms, ease of operation, simple fabricated structure, portability/miniaturization, simultaneous detection of multiple biomarkers, ability to perform onsite biodetection in a simple, rapid, sensitive, selective, and cost-effective manner, and their further extension to other relevant applications in a nonspecialist market.

This book aims to serve as a complete guide and an important source of information to all the stakeholders including scientists, engineers, and clinicians for augmenting their knowledge on the existing state of electrochemical biosensing for the detection and monitoring of food quality, adulterants, environmental pollutants, and disease diagnostics for point-of-care (POC) applications.

Chapter 1

Electrochemical biosensing: Progress and perspectives

1.1. Introduction

The current global trend of increased population density is impacting the environment alongside inappropriate antimicrobial use, the proliferation of multidrug-resistant pathogens, the emergence of new infectious agents, and ease of rapid disease dissemination are among the major challenges for the management of infectious diseases. Rapid and progressive development of nanotechnology has ensured the design and synthesis of novel electrochemical biosensors with improvised features (Campuzano et al., 2017c). Recent developments in the field of molecular biology, nanofabrication protocols, labeling, nano-instrumentation, multiplexing capabilities, along with microfluidics allows rapid, sensitive, selective, and easy-to-use electrochemical immunosensors that will find an important niche for detecting circulating biomarkers in the clinical field (Campuzano et al., 2017d).

The main focus of this chapter is to provide an updated critical overview of different recently reported nanomaterial-based electrochemical sensing and biosensing strategies based on various analytical approaches for the point-of-care (POC) healthcare-related diagnostics, evaluation of contaminants, additives and adulterants in foods, and environment management. Each section under the topic is discussed in its exhaustive details incorporating significant literature spanning over two decades.

1.2. Biosensors

Biosensors came into existence with the pioneering work of Leland Charles Clark Jr. in 1956, when he used an electrode for the detection of oxygen concentration in blood. Other scientists whose work have equally contributed to the surge in the area of biosensors are Clark and Lyons (1962), Guilbault et al. (1962), Updike and Hicks (1967), Guilbault and Montalvo Jr. (1969), and Di Gleria et al. (1986). After that, biosensor's design and diverse applications have seen a remarkable upsurge with biosensors now becoming the part of the modern lifestyle (Ali et al., 2020). For instance, glucose, lactate, and Scoville sensors are used to monitor blood glucose level, lactate, and chili respectively.

Electrochemical Biosensors. https://doi.org/10.1016/B978-0-323-90632-6.00003-0

Supermarkets are stocked with pregnancy kits employing paper-based lateral flow assays and electrical signal outputs. These are available under different trade names. Biosensors have wider application in the fields ranging from biomonitoring of critical disease biomarkers, biomonitoring of environmental pollutants, and food adulterants, etc. Sensors are available in four main categories, i.e., optical, acoustic, electrochemical, and mechanical (Rainbow et al., 2020).

A biosensor is equipped with an integrated receptor and a transducer device that converts the biorecognition element into a measurable physio-chemical output signal, which is proportional to the concentration of the target analyte. Selective interaction of bioreceptors with target analyte yields signal in the form of mass-change, heat, charge, and pH (Fig. 1; Bhalla et al., 2016). Most commonly used bioreceptors are enzymes, cells, antibodies, molecularly imprinted polymers (MIP) based molecular recognition, and nucleic acids. The functional aspect of biosensors involves interaction between bioactive agents, the transducer, and a signal processing unit. Biosensors have two main components, i.e., a target recognition element as receptors, nucleic acids, antibodies, and a signal transducer that converts target recognition into physically detectable signals (Ali et al., 2017; Hameed et al., 2018). Depending on the choice of transducers, biosensors can be of three types, i.e., optical, electrochemical, and mechanical (Fig. 2). Depending upon the bio-detection mechanism, biosensors can be of biocatalytic or affinity type. Biocatalytic sensors monitor the enzyme target reaction to produce electroactive molecules. Affinity sensors monitor the interaction between the bioreceptor and target analyte to generate a measurable signal (Ronkainen et al., 2010; Santos et al., 2012).

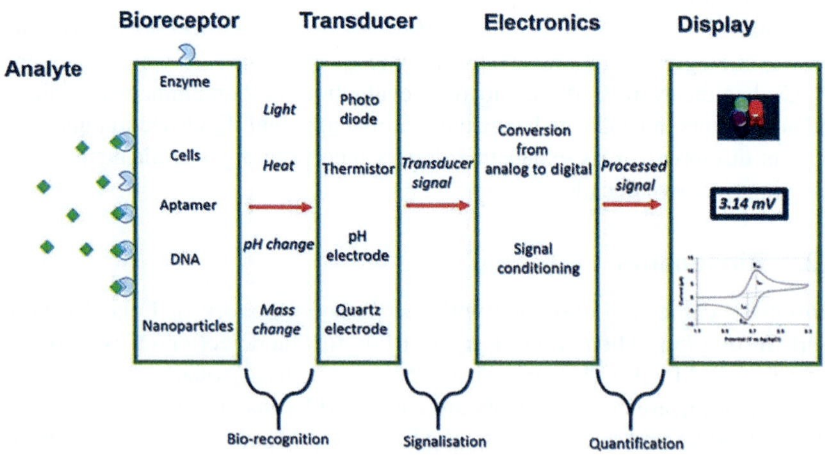

FIG. 1 Standard components of a biosensing device (Rainbow et al., 2020). *Reprinted from Bhalla, N., Jolly, P., Formisano, N., Estrela, P., 2016. Introduction to biosensors. Essays Biochem. 60, 1–8. Copyright © 2020 by the authors. Licensee MDPI, Basel, Switzerland. This article is an open-access article distributed under the terms and conditions of the Creative Commons Attribution (CC BY) license (http://creativecommons.org/licenses/by/4.0/).*

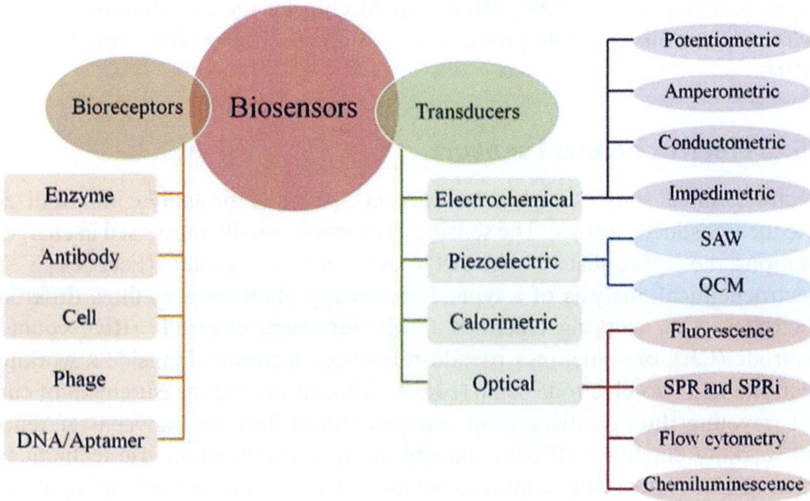

FIG. 2 Components and classification of the biosensor (Wu et al., 2019). *Copyright © 2019 by the authors. Licensee MDPI, Basel, Switzerland. This article is an open-access article distributed under the terms and conditions of the Creative Commons Attribution (CC BY) license (http://creativecommons.org/licenses/by/4.0/).*

1.3. Optical biosensors

These sensors are based on change in a fluorescence or color during or after the recognition event. The important transducer elements used for the fabrication of optical biosensors are internal reflection, fluorescence resonance energy transfer (FRET), chemiluminescence, bioluminescence, and surface plasmon resonance (SPR) (Sankarankutty, 2014). Principal optical techniques of optical biosensors are colorimetry, fluorescence, chemiluminescence, and surface plasmon resonance (SPR) (Alamer et al., 2018).

1.4. Mechanical biosensors

Mechanical biosensors measure a mass-sensitive surface deflection as target analyte bind to the functionalized surface. Based on the sensor-analyte chemical reaction, mechanical biosensors are classified into the following types, viz. affinity-based assays, fingerprint assays, separation-based assays, and spectrometric assays (Arlett et al., 2011). Quartz crystal microbalance (QCM) is a mechanical biosensor. It monitors the shift in the mass in sub-nanogram amounts. A QCM biosensor facilitates fast, sensitive, and cost-effective detection. A QCM biosensor recognizes the change in mass by the resonant frequency of the crystal. This technique has important application in the quantification of whole-cell organisms (Yu et al., 2018). A QCM-aptasensor was developed employing magnetic nanoparticles (NPs) and the QCM chip for the detection of *Brucella melitensis* in milk and milk products with high specificity (Bayramoglu et al., 2019).

1.5. Electrochemical sensors

Electrochemical sensors exploit the inherent charges of the analyte immobilized onto the transducer surface. The sensing platform is usually immersed in charged solutions, i.e., phosphate buffer (PB) and ferro/ferricyanide $[Fe(CN)_6]^{3-/4-}$. Electrochemical analysis of a typical biosensing platform uses three different electrodes, viz., working electrode (WE), reference electrode (RE), counter electrode (CE), or sometimes pseudo-reference electrode alongside a working electrode in a two-electrode setup (Fig. 3). This allows the measurement of current traveling through the system and potential difference between reference and working electrode utilizing amperometric and potentiometric techniques. These sensors are usually connected to the potentiostat to control the input signal and record the output signal. The output signal reflects the concentration of the analyte (Rainbow et al., 2020).

Electrochemical sensors have several advantages over their counterparts such as a lower limit of detection (LOD), a wide linear response range with many target analytes, superior self-assembled monolayer (SAM) stability, and high reproducibility. These are widely used for water analyses through handheld pH meter (potentiometric), integrated components in high-pressure liquid chromatography (HPLC) (amperometric), food adulteration, environmental

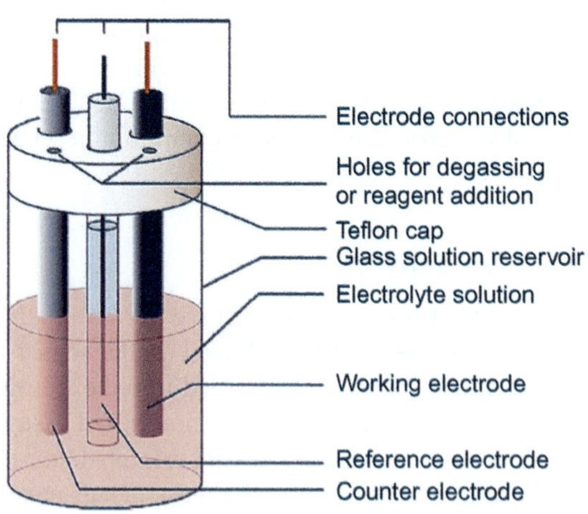

FIG. 3 Typical 3-electrode cell setup with a working electrode usually made of gold, a reference electrode, which may be Ag/AgCl, Hg/HgSO4, or SCE, and a thin-wire Pt counter electrode (Wu et al., 2019).

pollutant monitoring, endocrine disturbing compounds, illicit drugs, and heavy metal ions, etc. (Abdalla et al., 2019; Chinnappan et al., 2020; Ding and Lisak, 2019; Fekry et al., 2020; Fiel et al., 2019; Mayedwa et al., 2019).

Electrochemical biosensors offer affordable, reagent-less monitoring of biomarkers at different molecular levels in diverse samples in a real-time manner. This means that the procedure can be carried out even in resource-limited remote settings. These are advantageous, particularly for complex clinical samples, such as undiluted liquid biopsies, whole cells, and paraffin-embedded tissues, which are difficult to analyze using conventional techniques or electrochemical biosensing (Yáñez-Sedeño et al., 2019).

1.5.1 Components of electrochemical sensors: Electrodes, transducers, or detector device

The majority of the biochemical reactions occur at the working electrode of a transducer. The geometry and size of the working electrode plays important role in sensor performance. The basal part of the electrode or the main substrate that connects the transducer with the signal processing unit is made up of highly conductive material such as Au, Pt, Pd, glassy-carbon, etc. Gold owing to its inertness, corrosion, resistance, high conductivity, stability, and bioactivity is preferred. The top part of the basal part of the working electrode is usually made up of carbon nanotubes (CNTSs) or graphene nanoparticles or quantum dots (QD) to increase the surface area, improved conductivity, and electrocatalytic properties (Ma and Chen, 2015; Martynenko et al., 2017). The use of nanostructured electrodes has significantly improved the LOD (down to fM range) of biosensors (Rasheed and Sandhyarani, 2017). Carbon-based solid electrodes perform better than metal electrodes. Metal NPs, graphene, carbon nanofibers (CNF), molybdenum disulfide (MoS2), nanoporous Pt-Ru alloy, molecularly imprinted polymers (MIP), and Indium tin oxide (ITO) are also used to enhance the sensitivity of working electrodes (Eissa et al., 2014; He et al., 2017; Zhang et al., 2016, 2017).

1.5.2 Electrical interface

In an affinity biosensor, the interface is the sensing medium between the bulk of solution (or sample) and transducer surface where biological events take place. The surface of the working electrode establishes the link between an analyte and the working electrode. Surface alteration increases the stability of the biological system, thereby improving operational and storage stability. The surface modification involves functionalization by introducing the sites of a known functional groups to enable binding with NPs, bio-recognition elements, and target analytes. The surface functionalization can be achieved by covalent or non-covalent bonding (Sharma et al., 2016; Zhu and Zhang, 2015). The covalent functionalization alters the material structure of the electrode surface while non-covalent functionalization involves surface interactions through hydrogen bonding or van der Waal forces (Bai et al., 2009). Immobilization of nanocomposites is achieved by covalent binding, adsorption, cross-linking, and entrapment. The modification of the electrode surface depends on the material of electrode, biological entity, and electrode architecture (Sassolas et al., 2012).

1.5.3 Recognition receptors

This involves high-affinity biomolecules used for the selective binding of a target analyte. The majority of the electrochemical biosensors are based on high-affinity antigen-antibody interactions. Aptamers are the promising alternative as the recognition receptor to the high-affinity analyte. Engineered natural receptors such as antibody fragments, affirmer proteins, and artificial receptors are also used as recognition receptors with high affinity to the target analyte (Cieplak and Kutner, 2016; Ferrigno, 2016; Sharma et al., 2016).

1.5.4 Antibodies

These are used as highly selective analytical reagents for immune-based detection assays. Both polyclonal (pAb) and monoclonal (mAb) antibodies show remarkable selectivity for a wide range of target analytes. Polyclonal (pAb) antibody recognizes the multiple epitopes in the target analyte while monoclonal (mAb) antibody recognizes a single epitope in the target analyte. Immunoassays are combined with the electrochemical sensor to give rise to an electrochemical immuo-sensing platform. Electrochemical sensor-based immunoassays are available as standard (non-competitive) direct configuration, competitive direct or indirect formats (Vogiazi et al., 2019).

1.5.5 Aptamers

Aptamers are short single stranded DNA (ssDNA) or RNA molecules, which fold and bind to a ligand by complementary shape interactions resembling antibody-ligand binding affinity. Aptamers with high affinity and specificity can be selected in vivo for a wide range of molecules such as small chemicals, lipids, or proteins. Aptamers are synthesized by the systemic evolution of ligands by exponential enrichment (SELEX). Most of the materials used for designing aptasensors are carbon-based, i.e., single or multiwalled carbon nanotube (SWCNT, MWCNT), graphene oxide (GO), metal nanoparticles (MNPs), fluorescent quantum dots (QDs), and magnetic beads (Mustafa et al., 2017). These are preferred over natural receptors as detection tools in biosensing applications as they offer easy modification, reproducibility, stability, small size permitting high immobilization densities, reusable aptamer modified surface after regeneration, and low cost. Some aptamers yield important conformational changes upon target binding, facilitating high versatility in the design of electrochemical biosensors (Cunha et al., 2018). Aptamers can be easily integrated into electrochemical biosensing platforms due to their ease of modification with various functional groups. Aptamers can also be combined with electrochemical transducers to produce electric signals, and provide a simple, accurate, and inexpensive platform for diverse types of applications. However, optical readouts of aptameric-binding events require high precision, expensive instruments, and sophisticated numerical algorithms to interpret the data (Hayat and Marty, 2014; Rapini and Marrazza, 2017).

1.5.6 DNA, enzymes, and artificial receptors

Double-stranded DNA (dsDNA) interacts with small molecules via covalent (outer electrostatic binding, groove binding, intercalation) and non-covalent interactions (chemical modification of DNA constituents). Intercalation and groove binding are the two most common interactions between small molecules and dsDNA. This selective interaction helps in designing novel biosensing strategies (Oliveira Brett et al., 2007). Enzyme inhibition in the presence of a substrate generates biocatalytic biosensors when enzymes are used as a biorecognition element. Enzyme activation may lead to affinity biosensors when enzymes serve as a label (horseradish peroxidase and alkaline phosphatase) (Beasley et al., 1983; MacKintosh et al., 1990). Artificial enzymes such as DNAzymes (e.g., G-quadruplex-DNAzyme) enhance the sensitivity of biosensors. Recently HRP-mimicking DNAzyme is increasingly used as a biocatalytic label replacing natural enzymes in electrochemical affinity biosensors (Alizadeh et al., 2017). It shows high catalytic activity towards specific substrates with improved stability over natural enzymes. DNAzymes can be denatured and renatured several times without losing their catalytic activity.

1.6. Development and evolution of electrochemical sensors

Advances in material science and printing technology have enabled greater flexibility in the design and development of low-cost, reliable, and selective electrodes for the detection of specific electrolytes. The electrochemical techniques that have accelerated the development of evolved sensors are screen-printed electrodes (SPE), integrated nanomaterials, and electrochemical immunoassays. The hybrids incorporating the advantages of the above are also developed and applied for the low cost, simple, selective, and specific detection of clinical biomarkers, environmental pollutants, and food-adulterants.

1.6.1 Screen-printed electrodes (SPE)

Different types of materials such as carbon-based materials, plastics, and ceramic materials are coated with doping agents to enhance electron transfer.

These materials are also coated with selectivity agents to capture analytes of interest using screen-printing technology such as photolithography. The coating materials may include metal complexing molecules, immobilized enzymes, or affinity agents like antibodies. The coated electrodes are used as working electrodes (WE) in voltammetric analysis (Badihi-Mossberg et al., 2007; Renedo et al., 2007). Screen-printed electrodes (SPEs) can be modified with metals, enzymes, or affinity molecules (Renedo et al., 2007). SPEs coated with Au, Hg, Bi or doped with Au, Ag are used in stripping voltammetry (SV) to determine hazardous metals and chemicals in biological fluids with a detection limit of ng/ml. SPEs modified with enzymes are used to detect enzyme activity, pesticide metabolites, phenolic compounds, heavy metals, cholesterol, and glucose, etc. in biological matrices by immobilizing enzymes on a sol-gel matrix or conducting polymer (ng/ml). Antibody immobilized SPEs have been used in conjunction with amperometric and voltametric detection of antigen in biomatrices at pg/mL level (Renedo et al., 2007).

The performance of SPEs can be enhanced further by using interdigitated electrodes (IDE). These electrodes measure large current at lower potential. This is achieved by reducing the distance between the electrodes. The use of IDE working electrodes increases the surface area of the electrode. Sensor geometry can be modified by increasing the surface area of the counter electrode while maintaining the areas of the other two electrodes the same. This results in the reduction in the current density required for electrochemical reactions at the counter electrode (Jothimuthu et al., 2011).

1.6.2 Synthetic receptors

Synthetic receptors are designed in response to the natural biorecognition elements due to their good sensitivity, robustness, and detection range. These are endowed with switch on/off features to recognize and transduce in response to the external stimuli. Novel advances such as toggle switches, synthetic entities mimicking natural molecules, and gene networks facilitate the design of switchable functions of the sensing elements. Examples of related biosensing technologies are: synthetic cell-based biosensors; artificial liposomes; and bio-inspired synthetic molecules like biomimetics, molecular imprinting polymers (MIP), aptamers, peptide nucleic acids, and ribozymes. These biosensors find interesting applications in molecular sensing, biofuel production, waste degradation, and fine chemical production.

1.7. Portability and miniaturization: Microfluidics in electrochemical biosensors

Electrochemical methods are amenable to the development of portable instrumentation owing to their simplicity, low power requirement, and miniaturization ability. Microfluidics represents the integration of multiple technologies such as biosensing, nanotechnology, material engineering, and microsystems with micro-volumes (10^{-9} to 10^{-18} L), and microchannels. It represents a micro-total analysis system (μTAS), integrating multiple miniaturized analytical processes into a single monolithic device (Escarpa, 2014). Advancement in microfluidics requires improvising microfabrication techniques, electrokinetics, hydrodynamic flow, and electrochemical detection (López et al., 2016). Microfluidics offers interesting applications in the biomonitoring of food, environment, and disease biomarkers. However, currently, it is constrained by the requirement of complex matrix preparation of a biological sample and fabrication steps. These challenges can be resolved by integrating protein and DNA biomolecules into the microfluidic system (Arduini et al., 2016). The integration of electrochemical microfluidics, and cell culture techniques presents a novel biosensing platform for food analysis. This integrated system detects food contaminants (allergens, adulterants, pathogens, spurious chemicals, etc.) by monitoring the contaminants induced changes in cell morphology and cell metabolism (Jiang et al., 2016).

1.7.1 Lab-on-chip (LoC) devices

Electrochemical sensors can provide an integrated lab-on-chip (LoC) approach with a key focus on portability, miniaturization ability, and high sensitivity (Chiriacò et al., 2018; Choi et al., 2017). Point-of-care (POC) handheld device requires the integration of a number of components such as electrochemical sensing electrodes, microfluidics, signal connectivity, power sources, and other miniaturized elements (Jolly et al., 2019; Moschou and Tserepi, 2017). This integration into a single sensing platform imparts the ability of autonomous sample collection, preparation, testing, analysis, data readout, diagnostics, and reassured rapid detection (Land et al., 2018). However, additional features such as sample preparation, preconcentration/amplification, device connectivity, and antifouling, etc. are needed before such sensing device could achieve its full potential of wider applicability in diverse setups. (Aracil et al., 2015; Cheng et al., 2016; Jolly et al., 2015).

1.7.2 Lab-in-briefcase (LiB)

Environmental biomonitoring especially water monitoring in lower-middle-income group settlements requires a device with a simple operational mechanism, cost of manufacture, and convenient device integration. Lab-in-briefcase (LiB), which is a self-contained portable system fulfills most of these requirements. It is equipped with elements required for water sample collection, sample pre-treatment, biosensing elements, and data processing unit. It is composed of a disposable sensor chips and reusable sensing devices thus reducing the overall cost. The sensing device is usually an electrochemical cell connected to a potentiostat. Signal collection and data processing can be performed by commercially available USB-powered potentiostats (Rainbow et al., 2020). The option for single-use, low-cost biochips are also available. Currently available options for electrochemical disposable biochips are microfabricated electrodes on various substrates such as screen-printed electrodes (SPEs); printed circuit boards (PCBs); and evaporated metals on glass slides, i.e., Au, Ag, etc. However, evaporated glass electrodes on glass slides lack mechanical strength and are susceptible to physical abrasion. Similarly, another major drawback of SPEs and PCBs are the lack of reproducibility and purity of metal surface during the commercial printing procedure (Smith et al., 2014). These platforms are easy to prepare in a microfabrication lab and can be custom-made through mass-production avenues (Moschou and Tserepi, 2017).

1.7.3 Advantage of microfluidics integrated with electrochemical biosensor

Microfluidics integrated with electrochemical biosensor enables easy sample preparation before analyte detection. This is particularly advantageous in the case of environmental water samples, which have a more complex matrix as compared to other samples such as blood, saliva, urine, sweat. Water samples vary with the catchment area and upstream activities such as urban settlements, agricultural activities, and industrial plants. Target analytes may be present in very low concentrations in the background of non-target analytes. Therefore, the device for POC applications should be equipped with the provisions of mixing, filtration, lysis/extraction, followed by sample preconcentration steps (Rainbow et al., 2020). Lee et al. (2012) have developed a microfluidic-based electrochemical sensing platform for fully automated mixing of various solutions including buffer, antibody conjugated nanomaterials, and target antigen. This platform was embedded with a gold-patterned microarray chamber for sandwich complex-based sensing and automated detection of multiple analytes in an hour. Another microfluidic-based electrochemical sensing platform utilized magnetism and magnetic beads. Mixing within the microfluidic device is achieved by rotating magnetic disks and magnetic beads (Shanko et al., 2019). Replicating the similar protocol, Owen et al. (2013) have achieved the mixing of complex analytes within a microfluidic device. Owen's group has shown that efficient sample mixing depends on the variables such as flow rate, rotation speed, and bead density.

Microfluidics-based sample preparation utilizes magnetic beads for sample preparation in on-chip devices. This ensures filtration of target analytes. These beads can be functionalized with analyte complementary molecules (antigens, antibodies) to specifically target analytes of interest and remove non-target components. This is especially advantageous for real-world natural samples having complex matrices. Zhuo et al. (2010) have functionalized the magnetic beads with the target-complementary molecules to capture target analytes from a complex sample matrix. The target was later separated by a magnetic-fractionator system. The microfluidic electrophoresis protocol within on-chip microfluidic devices is increasingly gaining traction for sample preparation. It offers advantages in terms of low-sample requirements, cost-efficiency, rapid detection, and easy integration with sensing modules. Various electrophoretic methods that have been incorporated in microfluidic devices are gel electrophoresis, capillary electrophoresis, and dielectrophoresis (Ou et al., 2019; Yu et al., 2016; Zhang et al., 2019).

1.8. Types of electrochemical biosensors based on electric signals

Immunosensors are a type of biosensors that detects real-time interaction kinetics between an antibody and antigen on a transducer surface. Antibody or antigen can be the species immobilized on the transducer surface to detect antigen or antibody, respectively. Electrochemical immunosensor analyzes electrical signal (voltammetric, potentiometric, conductometric, or impedimetric) produced on an electrochemical transductor (Fig. 4). Electrochemical immunosensors are advantageous owing to high specificity, selectivity, portability, and ease to carry out in situ or automated detection. Electrochemical sensors find important applications in environmental analysis, detection of additives or contaminants in food, and as a biomarker for disease diagnosis like cancer (Felix and Angnes, 2018).

1.8.1 Amperometric sensors

The Amperometric technique measures the current related to the redox potential of molecular species such as redox-capable ions in a solution. Flow-injection analysis (FIA) or sequential-injection analysis (SIA) coupled with amperometric detection or micro-electrical mechanical systems are useful for the determination of toxic substances (Chen et al., 2006; Gilbert and Sale, 2005; Neufeld et al., 2000; Sole and Alegret, 2001; Wang et al., 2001; Xu et al., 2007). Amperometric detection integrated with micro-electrical mechanical systems was utilized for the detection of residues in chemical nerve agents, nitroaromatic compounds, and heavy metals (Dabek-Zlotorzynska et al., 2003; Wang et al., 2001, 2002). Flow injection analysis (FIA) coupled with amperometric detection facilitates the real-time, continuous-flow detection of toxic compounds in environmental and biological samples utilizing very small sample volumes ($<20\,\mu l$) with high-throughput analysis of samples (3600 injections/hour) (Liu et al., 2007; Liu and Lin, 2005). Amperometric detection is a superior technique in the flow-electrode system or with a rotating or vibrating electrode (Heineman et al., 2018). Amperometry is one of the most suitable technique in mass-transfer conditions as it minimizes the background signal from the charging current that would interfere with the electrochemical quantifications. It monitors enzyme reactions in a biocatalytic sensor (Vogiazi et al.,

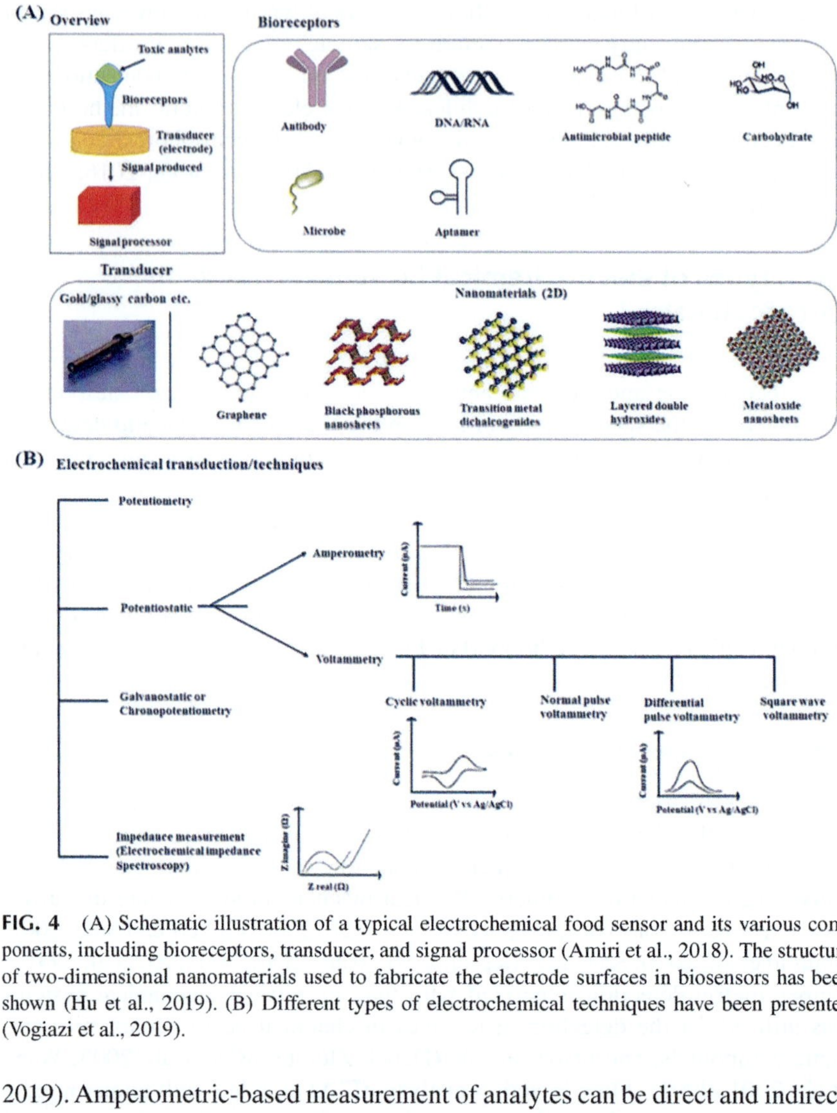

FIG. 4 (A) Schematic illustration of a typical electrochemical food sensor and its various components, including bioreceptors, transducer, and signal processor (Amiri et al., 2018). The structure of two-dimensional nanomaterials used to fabricate the electrode surfaces in biosensors has been shown (Hu et al., 2019). (B) Different types of electrochemical techniques have been presented (Vogiazi et al., 2019).

2019). Amperometric-based measurement of analytes can be direct and indirect. Direct amperometry works based on the relationship between the measured current and the products of the redox reaction. Indirect amperometry measures the analytes' metabolic substrate or product using conventional detectors (Gerard et al., 2002).

1.8.2 Amperometric immunosensors

Amperometric immunosensors measure the changes in current, which is directly proportional to the analyte concentration, at the constant voltage indicating the redox reaction on the analyte. By labeling the electroactive materials like enzymes, antigens, or antibodies, their concentration can be determined by monitoring the current change. The current signal resulting due to antigen-antibody binding can be amplified by the markers such as alkaline phosphatase (ALP), horseradish peroxidase (HRP) or glucose oxidase or nanomaterials (Au nanoparticles (AuNPs), graphene, carbon nanotubes (CNTs) or silica nanoparticles (Yáñez-Sedeño et al., 2019; Aizawa et al. (1979). An amperometric immunosensor was developed for the detection of chorionic gonadotrophin for the first time in 1979. Amperometric immunosensing platforms can be of two types, sandwich-type, and competitive type. An amperometric immunosensor was developed, where rabbit antihuman CD9 antibodies were immobilized on Au-substrate using mAb against CD9 to detect capture antigen. This type of immunosensor can be integrated with a portable, semi-automatic system for detection. Manfredi et al. (2016) have first reported a competitive type amperometric immunosensor using gliadin-functionalized carbon/nanogold-SPE for rapid detection of celiotoxic prolamins (Yáñez-Sedeño et al., 2019).

1.8.3 Potentiometric sensors

The potentiometric technique measures the potential difference between the reference and working electrode. This potential difference is precisely controlled to investigate the behavior of ions in the solution affected by biomolecules on the transducer surface under the condition of zero or little current. This gives the ion-activity data at the interface. Both voltammetry and amperometry provide quantitative information on current with respect to time, but voltammetry is advantageous as it provides current response information for reversible reactions as well by pulsing an applied potential between high and low values. If the potential is pulsed between high and low values with a short time interval, the current that flows through the electrode can be measured within the lifetime of the pulse. Due to high and low pulse in voltammetry, the analyte at the electrode surface undergoes a redox reaction generating positive or negative current flow. A plot between the difference of changing the current flow and potential yields a voltammogram, which is characteristic of the electroactive species (Skoog and Leary, 1992). Potential applied in a staircase fashion produce waveform as a potential per unit time, the technique known as square-wave voltammetry (SWV). Stripping analysis involves preconcentrating or electrodepositing a solute on the electrode surface using a constant applied potential before the application of stripping potential, a technique known as stripping voltammetry (SV). Square-wave stripping voltammetry (SWSV) occurs when stripping potential is applied as a square wave. Voltammetry and amperometry offer the detection limit within the range of 10^{-7} to 10^{-8} while stripping voltammetry with solute preconcentration step gives the detection limit within the range of 10^{-10} to 10^{-11} (Bard and Faulkner, 1980) (Fig. 5).

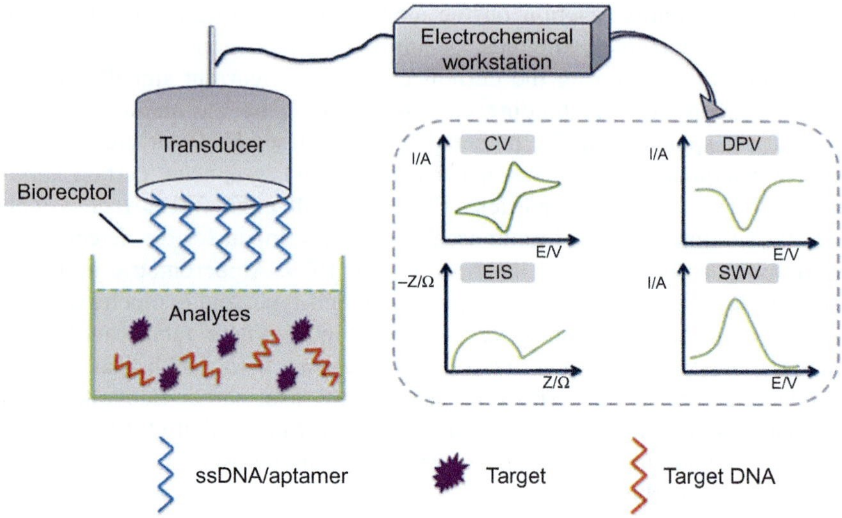

FIG. 5 Schematic diagram of electrochemical DNA biosensors (Wu et al., 2019). *Copyright © 2019 by the authors. Licensee MDPI, Basel, Switzerland. This article is an open-access article distributed under the terms and conditions of the Creative Commons Attribution (CC BY) license (http://creativecommons.org/licenses/by/4.0/).*

Other voltammetric techniques used in the electrochemical biosensors are cyclic voltammetry (CV), normal pulse voltammetry (NPV), and differential pulse voltammetry (DPV). CV is the most versatile voltammetric technique to study electroactive species. CV experiments are consisting of scanning from an initial fixed potential to a switching potential at which scan direction is reversed towards a final potential, where the resulting current is measured. This technique may involve single scans as well as multiple scans (Heineman et al., 2018). The potential pulses are applied to the working electrode to improve the LOD by minimizing the charging current associated with changing electrode potential. NPV consists of series of potential pulses of increasing amplitude for which the current is measured near the end of each pulse after the interfering charging current is decayed away. DPV involves the application of a constant amplitude pulse that is scanned through the fixed potential range. In the DPV technique, current is measured before the pulse is applied and at the end of the pulse. The difference in the two currents is displayed in the voltammogram. SWV is preferred over NPV and DPV as it employs faster scan rates than both NPV and DPV and offers lowers LOD. SWSV offers enhanced sensitivity and lower LOD (Kissinger and Ridgway, 2019; Wang, 2018) (Fig. 5).

1.8.4 Potentiometric immunosensors

Potentiometric immunosensors perform real-time monitoring of potential change for the direct or indirect determination of antigen. There is a logarithmic relationship between the potential change and antigen or antibody concentration. Potentiometric immunosensors can be of two types depending on the measurement of membrane or ion-electrode potential. Measurement of membrane potential is fraught with non-specific adsorption issues leading to large background current and lower detection sensitivity, while ion-electrode potential measurement results in highly sensitive and selective measurement. Ghindilis et al. (1992) have reported the first ion-selective immunosensor for the detection of insulin using lactase labeled antibody for insulin detection. Fu et al. (2005) have synthesized AuNP incorporated ion-selective immunosensor where AuNP was adsorbed on the electrode surface by self-assembly of mercaptoethylamine followed by immobilizing the IgG to develop a potentiometric immunosensor. Tang et al. (2004) have designed a similar sensor but used a glassy carbon electrode (GCE) to immobilized anti-diphtheria IgG for the detection of diphtheria toxoid in biological samples.

1.8.5 Impedance sensor

Impedimetric technique measures the complex resistance and capacitance related redox state of molecular species. An impedimetric sensor conducts a low amplitude wide-range AC frequency scan to record complex impedance. This is further analyzed by modeling the impedance data with an ideal electric circuit such as Randles equivalent circuit. This is done to obtain charge transfer resistance (Rct) in a Faradic process with a redox couple or, by converting the impedance into capacitance to investigate the intrinsic property of the electrochemical double layer at the interface (non-Faradic impedance) (Barry et al., 2009).

Electrochemical impedance spectroscopy (EIS) is a powerful technique in electrochemical affinity biosensors as it offers simple and label-free measurement of analytes. In EIS, an alternating potential is applied to the electrochemical cell whose frequency varies over a wide range to obtain an impedance spectrum. The resistive and capacitive components of impedance provide important information to study interface properties and surface reactions. Variations in the electron-transfer resistance at the working electrode give direct monitoring of the analyte binding to a recognition element such as antibody or aptamer. Aptasensors are label-free impedimetric immunosensors used to detect small molecules in environmental samples (Kissinger and Ridgway, 2019). EIS biosensors have interesting applications in food monitoring and environmental monitoring.

EIS is a robust technique performing the detection of an analyte in a label-free manner. EIS finds significant application in food industry biomonitoring including, ripening, moisture content, the distinction between fresh and frozen-thawed foodstuffs, additives in juices, and adulterants (Zhao et al., 2017).

1.8.6 Impedance immunosensors

Impedance immunosensors combine the benefits of electrochemical impedance spectroscopy such as capacitance, inductance, and resistance to produce a specific impedance signal, which is largely utilized to study the properties of conductive materials and interfaces. These sensors monitor the label-free, real-time change in the transfer of electrons in solutions before and after an immune reaction to determine the test analyte. The design of these sensors use nanomaterials for antibody immobilization, their maintenance, and activity over a long period (Kharitonov et al., 2000). Chen et al. (2008) have reported an enzyme-free impedance immunosensor using 3D-ordered macroporous gold film for the immobilization of antibodies to detect C-reactive protein (CRP). A sandwich immunosensor using CNT and AuNP (as electrode matrix for antibody immobilization) was constructed for the detection of glycoprotein (Zhuo et al., 2010). A marker-free microelectrode impedance sensor was designed for the detection of *Escherichia coli* O157:H7 in food samples. The test bacteria were adsorbed on Indium Tin oxide (ITO) microelectrode leading to impedance response directly proportional to the concentration of test bacteria. Commercialization of EIS biosensors is constrained by several factors such as the complicated process of impedance detection, stability of biomolecule immobilization, smaller analytes, and susceptibility to non-specific adsorption (Neethirajan et al., 2018).

1.8.7 Conductometric sensors

The conductometric technique measures the change of conductance for a solution after recognition of charged target analyte on the biorecognition event. These sensors monitor the changes in the number and type of ions in the solution before and after the immune reaction to detect an antigen. The quantification involves immobilizing a specific enzyme on the surface of the metal electrode to determine the conductivity under the action of an electric field. Yagiuda et al. (1996) have designed this type of immunosensor for the detection of methamphetamine in human urine in a cost-effective manner. While Muhammad-Tahir and Alocilja (2003) have used this sensor for the detection of *Escherichia coli* O157:H7 and *Salmonella spp* in food samples. This type of immunosensor was also utilized for the detection of alpha-fetoprotein (AFP) and polypeptide antigen (Tang et al., 2011; Xu, 2016). These sensors have few limitations like the large influence of ionic strength in test samples, non-specific adsorption, and buffer volume in the measurement process thus limiting its use.

1.8.8 Capacitive sensors

Electrochemical capacitive immunosensors monitor the changes in the capacitance of the interface between an electrode and testing solution. Typically, when biomacromolecules like proteins with low polarity get adsorbed onto the electrode surface, the thickness of the interface layer increases while the di-electric constant of the interface layer reduces resulting in reduced capacitance, reflecting the amount of antigen detected. Mirsky et al. (1997) have pioneered the design of this type of sensor for the detection of albumin in human serum using SAM for the immobilization of protein molecules.

1.8.9 Gravimetric sensors

Gravimetric biosensors offer highly sensitive label-free detection of target analytes. These are piezoelectric devices such as quartz crystal microbalance (QCM) and micro- (or nano-) mechanical cantilever setups that detect a biological target that depends on the change in resonance frequency proportional to the mass of the analyte (Meisam et al., 2013).

1.9. Nanomaterials for electrochemical sensor applications

Nanomaterials such as colloidal gold/silver semiconductor NPs, and markers-loaded nanocarriers (carbon nanotubes, apoferritin, silica nanoparticles, and liposome beads) play a significant role in the improvisation of electrochemical sensors and immunosensors. Nanomaterial-incorporated sensors lead to enormous signal enhancement due to the formation of nanomaterial-antibody-antigen complex. This forms the basis of ultrasensitive detection of protein biomarkers related to disease, biothreat agents, and infectious agents. However, desired targets to achieve are signal amplification of various orders of magnitude, lower detection limit, and simultaneous detection of multiple targets (Liu and Lin, 2007). Nanomaterials are highly selective and can assay compounds having similar structures but different electrochemical behavior. These can be used in different structural configurations and oxidation states. These can be functionalized with organic and inorganic molecules coupled with other polymeric structures or other nanomaterials (hybrid NMs) (Blasco et al., 2007; Blandón-Naranjo et al., 2018b, 2018a).

1.9.1 Nanohybrids

Electrochemical biosensors have been greatly enriched by the outstanding conductivity and biocompatibility of **2D-NMs and 2D-nanohybrids**. Clinical biomarkers based on veritable hybrid 2D-NMs are available particularly hybrid nanostructures involving graphene and other graphene-like 2D-layered nanomaterials (GLNs). Different strategies have been optimized for utilizing 2D-nanomaterial hybrids/heterostructures for developing immunosensors for tagging or modifying electrode transducers called hybrid nanocomposites (Campuzano et al., 2017c, 2017d). However, electrochemical immunosensors comes with a limitation in terms of instability, volatility, non-reproducibility and other shortcomings of biological identification unit like antibody (Yáñez-Sedeño et al., 2019) (Fig. 6).

2D-nanomaterials and 2D-based nanohybrids show excellent conductivity and biocompatibility for their use in electrochemical biosensors. Hybrid nanostructures involving graphene and other graphene-like 2D-layered nanomaterials (GLNs) are particularly highly advantageous (Campuzano et al., 2017a, 2017b, 2017c, 2017d). 3D-graphene materials particularly free-standing porous

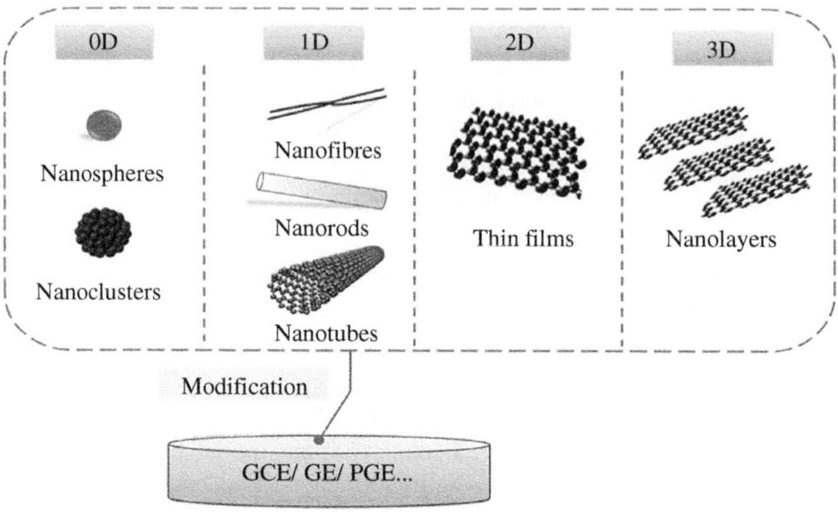

FIG. 6 Nanomaterials commonly used in the modification of electrochemical DNA biosensors (Wu et al., 2019). *Copyright © 2019 by the authors. Licensee MDPI, Basel, Switzerland. This article is an open-access article distributed under the terms and conditions of the Creative Commons Attribution (CC BY) license (http://creativecommons.org/licenses/by/4.0/).*

graphene with tunable pore size and void space shows great potential in biosensing and bioelectronics. They come up with additional advantages like interconnected porous network, a high electronic conductivity, a large active surface area, and good chemical/thermal stability thus putting them over dispersed graphene sheets and make them good candidates for high-performance electrode materials in bioelectronics devices like biosensors and biofuel cells (Qiu et al., 2017). Graphene-based composite materials have a major role in sensor platforms due to their atom-thick 2D-conjugated structures, high conductivity, large specific surface areas, and controlled modification. GCFs show unique optical, mechanical, electrical, chemical, and catalytic properties resulting in sensors with high sensitivity, selectivity, and stability (Yu et al., 2017).

Electrochemical sensors based on 2D-graphene materials like boron nitride (BN), graphite-carbon nitride (g-C_3N_4), transition metal dichalcogenides (TMDs), transition metal oxides, and graphene are also available. These are better suited for the detection of important biomarkers like glucose, hydrogen peroxide, and cancer biomarkers (Bollella et al., 2017). 2D-graphitic carbon nitride nanosheets (CNNSs) with planar graphene-like structures have superior stability, high fluorescence quantum yield, low toxicity, excellent biocompatibility, unique electroluminescent and photoelectrochemical properties that make them suitable candidates for biosensing. CNNSs have great utility in photoelectrochemical biosensing, electrochemiluminescence biosensing, and fluorescence biosensing. Integration with nucleic acid makes them a good candidate for intracellular imaging (Xiong et al., 2017).

1.9.2 Nanoparticles (NPs)

Nanoparticles (NPs) constitute an important component of a biosensing platform as almost any material can be shaped into nanosized structures thus imparting specific properties to the sensing elements. Noble metals like Au and Ag induce LSPR leading to an enhanced SPR or Raman signals (El-Ansary and Faddah, 2010; Doria et al., 2012). NPs aid in the ultrasensitive SPR detection of small molecules by enhancing the SPR signals and pre-concentrate low levels of target analytes in the sample. NPs are widely utilized as high mass labels in SPR assays. They enhance the binding signals and also the detection sensitivity, which is largely governed by the size and distance of the nanoparticles from the sensing surface. Signal enhancement increases in a proportionate manner with the size of the NPs, and reduced distance of NPs from the sensing surface. However, using large NPs (≥ 100 nm) for SPR detection poses a serious challenge as the short effective penetration depth of the SPR evanescent field, and steric hindrance of binding when small molecules are immobilized on the surface (Wang et al., 2019).

Semiconductor nanomaterials like silica NPs and quantum dots (QDs) exhibit large surface area, high surface reaction activity, strong adsorption, careen for immobilization of proteins and enzymes, easy functionalization due to a large number of unsaturated groups and hydroxyl groups, etc. facilitating their use in biosensor design for enhanced sensitivity (Liu et al., 2011). Silica NPs can be utilized for bio-imaging, optical bioassays, drug delivery, diagnosis, and therapeutic applications (Liu and Lin, 2007). Semiconductor NPs such as quantum dot (QD), CdS, PbS, and ZnS owing to their unique optical properties like size-tunable fluorescence and sensitivity play important role in biosensing. **Quantum Dots (QDs)** can be used as electrochemical and fluorescent labels (Yáñez-Sedeño et al., 2019). QDs are semiconducting nanocrystals required for fluorescence-based transduction in DNA or aptamer-based sensing. QDs availability in different wavelengths makes them interesting candidate for multiplexed analysis (Algar et al., 2017). QDs can be combined with AuNPs to increase the excitation rate of QDs and fluorescence, and with magnetic NPs for improved bio-detection (Moro et al., 2017).

1.9.3 Carbon-based nanomaterials

Carbon-Based Nanomaterials such as carbon nanotubes (CNTs), graphene, graphene oxide (GO), carbon nanofibers, carbon nanospheres, bucky-paper, etc. offers good conductivity, biocompatibility, robust mechanical strength, large surface area for an increased amount of immobilized biomolecules, and improved active sites for electrochemical reactions (Yáñez-Sedeño et al., 2019). The ability of carbon to hybridize sp, sp2, and sp3 configurations with narrow gaps between their 2s and 2p orbitals impart versatility in the design of carbon-based nanomaterials for the sensitive detection of biological analytes (Wanekaya, 2011; Singh, 2014, 2016a, 2016b).

Carbon nanotubes (CNTs) have recently seen a spurt in their use for building electrochemical sensors/biosensors and to study the electron transfer capabilities of proteins. CNTs offer a large surface area to weight ratio, and good electrical conductivity, finding important applications in the fabrication of immunosensors (Wang et al., 2004; Zhu et al., 2005; Yu et al., 2006). Carbon nanotubes (CNTs) are equipped with unique structural, mechanical, and electronic properties permitting their role in different sensing protocols.

Graphene exhibits unique physiochemical properties like a monolithic structure, high conductivity, large specific surface area, no toxicity, and good electron mobility facilitating its use in electrochemical biosensing (Yáñez-Sedeño et al., 2019). Graphene is a carbon material, where a 2D-sheet of carbon is linked by sp2 bonds. This configuration endows graphene with extraordinary properties such as high conductivity and electron mobility at room temperature, low energy dynamics of electrons with atomic thickness, robust mechanical behavior, and flexibility making it a boon for electrochemical and analytical applications (Felix and Angnes, 2018).

C60 Fullerenes and Carbon Dots or graphene quantum dots/C60 is a non-dimensional carbon nano-object utilized as a redox mediator and nano-scaffolds in enzymatic sensors. It can be combined with AuNPs and PtNPs for improved electrochemical, gravimetric, immune, and electroluminescent aptasensing applications (Zhuo et al., 2014). Carbon QDs are equally efficient as semiconductor QDs regarding FRET-based biosensing and as fluorescence labels (Bu et al., 2014). AuNP and carbon QDs modified with antibody-antigen system leads to efficient FRET between them (Shi et al., 2014). Recently, the field of analytical chemistry has witnessed a fillip in the use of quantum dots (QDs) as nanomaterials.

Carbon nanofibers are equipped with a large functional surface area and more surface-active groups, while carbon nanospheres have rich functional groups imparting good biocompatibility, dispersibility, and activity. Soluble carbon fibers and carbon nanospheres are used to construct highly sensitive immunosensor (Yáñez-Sedeño et al., 2019). For electrochemical biosensing, glassy carbon, doped diamond, and graphite are standard materials for electrodes. Their nanostructured part in the form of **carbon nanotubes fullerenes**, or graphene partly became the material of choice for improved performances of bioanalytical devices. Recently, **carbon nanodots** have replaced QDs as a promising alternative for biosensing and imaging. For electrochemical transduction, carbon paste electrodes (CPE) may combine with any carbon nanomaterials for improved biosensing (Holzinger et al., 2017). **Carbon buckypaper** occurs as a thin film (5–25 μm), consists of laterally arranged networks of nanotubes used for synthesizing nanocomposites. It has been utilized for the construction of a mediator-free glucose sensor with high sensitivity, stability, selectivity, and reproducibility (Ahmadalinezhad et al., 2011). **Carbon black** is a carbon material extensively utilized as filler for rubber, plastic, and composites preparation. Its primary structure is composed of spherical particles while the secondary structure is composed of aggregates. CB dispersion followed by sonication in appropriate solvents leads to carbon nanoparticle (CNPs) formation. Nano-CB exhibits excellent conductivity and electrochemical properties.

1.9.4 Apoferritin nano-vehicles and metal phosphate labels

Apoferritin is also used as label. These are easy to prepare and functionalize requiring only simple diffusion and pH adjustment. Metal phosphate NP labels offer a lower detection limit than semiconductor NP labels. **Liposomes** serve as a preferred bioassay label owing to high surface area, large internal volume, and ability to get linked with a variety of bio-recognition elements via its lipid layer. The label may incorporate enzymes, fluorescent dyes, electrochemical and chemiluminescent markers, DNA, RNA, ions, and radioactive isotopes. These labels could be either ion-encapsulated liposome labels or ion-selective electrodes for serum protein analysis. Liposomes encapsulating electroactive materials like ferrocyanide and ascorbic acid are used as labels for electro-immunosensing applications (Liu and Lin, 2007; Singh, 2017a).

QDs are semiconductor nanocrystals that come up with size-dependent fluorescent properties. QDs score over conventional organic fluorescence probes having the advantages of broadband excitation, narrow bandwidth, and high-intensity emission. QDs have important applications as fluorophores in biological applications due to the size-tuned fluorescent emission as a function of the core size, shape, and material. **Magnetic microspheres** are magnetic-nanomaterials displaying magnetic response and nano-synergistic effects along with exhibiting characteristic features of magnetic microspheres. Other nanomaterials of choice could be **Magnetic gold nanospheres**, **NPs-functionalized phosphopeptide nanospheres**.

Other labels of interest might be polystyrene beads and microcrystals (ferrocenemicrocrystal), sub-micrometer metallic microrods, metal, or alloy nanowires. These can be utilized for product identification, electrochemical DNA hybridization assay, and optical bioassay. Nanowires are metallic and non-metallic 1D-nanostructure. Nanowire-conjugated enzymes are explored for developing biosensing devices and electrocatalytic applications due to their high aspect ratio, high surface-to-volume ratio, low or no toxicity, and biocompatibility (Bagal-Kestwal et al., 2018). **Upconverting Nanoparticles** (UNCPs) can replace fluorescent labels for biosensing applications due to low background emission, high excitation wavelength, and lack of absorption effects with other components of the biosensor. The use of UNCP for biosensing has one limitation, i.e., the requirement of 980 nm of wavelength which lies in the absorption range of water. This might raise the temperature of the sample thus interfering with the bioimaging process. However, the use of neodymium lowers the excitation wavelength to 808 nm (Wang et al., 2013a, 2013b; Himmelstoß et al., 2017).

1.10. Electrochemical immunoassays

Electrochemical immunosensors integrates traditional immunoassay with modern biosensors and electrochemical analysis, thus incorporating high selectivity of immunoassays and high sensitivity of the electrochemical analysis. The electrochemical signals can be electric current, potential, conductance, and impedance. The working of immunosensors depends on the specific recognition event between antigen and antibody. The biological molecules such as proteins, hormones, antibiotics, persistent environmental pollutants, and toxins acting as haptens triggers the release of specific antibodies ensuring high selectivity of the reaction. Immunosensors can be of electrochemical, optical, piezoelectric, or calorimetric types, etc. (Singh, 2014, 2016a, 2016b; Pan et al., 2017).

Electrochemical immunosensors and immunoassays have gained considerable attention among the scientific fraternity along with industries owing to high sensitivity, low cost, and inherent portability (Ghindilis et al., 1998; Warsinke et al., 2000; Stradiotto et al., 2003; Lin and Ju, 2005; Bakker and Qin, 2006; Marquette and Blum, 2006). Electrochemical immunosensors and immunoassays work on the principle of highly specific antibody-antigen interaction, where immunosensor work by immobilizing the immunologic material onto the electrochemical transducer while immunoassay work by immobilizing the immunologic material on a solid support such as microplate wells, magnetic beads, and polystyrene beads. Nanomaterials (NMs) are attached to the transducer surface or solid support following immunoreactions. Quantification involves measuring the specific activity of an NM after the release step or after adding a substrate. Transducing signals of antibody-antigen reaction can be amplified using labels like enzymes, electroactive molecules, redox complexes, and metal ions (Liu and Lin, 2007; Singh, 2017a, 2017b).

Electrochemical immunoassays can be of (non-competitive) direct configuration, competitive direct, or indirect formats. The transducer's probe consisting of a microarray of immobilized antibodies on its surface allows standard (non-competitive) direct assays, competitive direct micro-assays, or sandwich assays as the preferred assay methods. Standard assays involve the direct binding of the target analyte to the immobilized antibody. The accumulation of the target on the electrode surface causes a change in the observed electrochemical signal. In competitive direct immunoassays, an unlabeled target molecule in a sample competes with the labeled target molecule to bind antibody on the electrode surface. The labeled but unbound analyte is separated by washing and the remaining labeled bound analyte on the biosensor surface is measured. The competitive indirect format utilizes an enzyme-labeled secondary antibody. Label-free protocols minimize the number of analytical steps paving for rapid detection of target analyte (Rogers, 2000). Antibody fragments such as Fab, Fv, and scFv owing to their small size, and high specificity to the target increases the selectivity and sensitivity in electrochemical affinity biosensing applications. The antibody fragments minimize non-specific binding and reduce steric hindrance. The functional efficacy of the antibody fragments increases when it is conjugated to nanoparticles (Wang et al., 2013a, 2013b).

1.10.1 Nanomaterials in electrochemical immunosensors

The advent of novel functional nanomaterials (NMs) and analytical technologies have greatly advanced the electrochemical immunosensor platforms/devices. The improvised immunosensor platform finds a broad application in the biological, biomedical, biotechnological, clinical, and medical diagnostics, environmental and health monitoring, and food industries. NMs come up with unique surface properties, small size, and macroscopic quantum tunneling along with unique mechanical, electrical, optical, magnetic, and catalytic properties. Structural properties like strong adsorption capacity, reliable orientation, biocompatibility, and structural compatibility facilitate these novel nanomaterials in the immobilization of biomolecules and play a crucial role in the development of immunosensors (Yáñez-Sedeño et al., 2019). The introduction of a wide array of NMs coupled with surface modifications and microfabrication techniques provide an opportunity for in vivo and in vitro clinical analysis via a specifically planned electrode/solution interface. NM incorporated sensing platforms offer the advantage of low cost, portability, energy efficiency, ease of fabrication, online monitoring, and simultaneous sensing capability in addition to high sensitivity, selectivity, and durability. These features provide the main thrust for the rapid growth, advancement, and continued improvisation in electrochemical immunosensing platforms incorporating expertise from the diverse fields of chemistry, biology, material science, and medicine (Singh, 2014; Maduraiveeran et al., 2018). The NMs based electrochemical immunosensor falls into two distinct categories, (1) NM-modified electrochemical transducers (2) NM-bimolecular conjugates as labels. Labeling of antibody or antigen with NM label does not interfere with its bioactivity (Liu et al., 2007; Wang, 2007).

1.11. Conclusions and future perspectives

Biosensors have seen significant progress since the time Clark and Lyon first devised a glucose sensor. Since then, there is a remarkable development in fabrication and its wider application that has encompassed almost every walk of human life. Among different types of sensors, electrochemical sensors have particularly made rapid and impressive advances with the changing needs of daily life. Electrochemical biosensors have interesting applications owing to their small size, portability, fast measurement, low sample size, simplicity, high sensitivity, and low cost. An overview of progress in analytical chemistry reaffirms that electrochemical sensors represent the most rapidly growing class of chemical sensors.

Nanobiosensors incorporates nanomaterials with an electrochemical sensor. The use of nanomaterials facilitates novel highly sensitive assays and improvised nanomaterials incorporated electrochemical/biosensing instruments. Nanoparticles enhance the response time of the sensor by improving the catalytic effect and increased electron transfer on the electrode, thereby improving the accuracy of sensor output. Nanoparticles perform 5-important functions in the sensing architecture of electrochemical biosensors such as receiver stabilization, catalysis, improved electron transfer, labeling, and participation in detection reactions. Other advantages of nanomaterials in the fabrication of sensing platforms are direct bonding of the enzyme to the electrode surface, amplification of the diagnostic signal, biocompatibility, and high adsorption power. However, further improvisations are required in the design of electrochemical biosensors to make them more user-friendly for their wider applications in a non-specialist market. A model biosensor should have high specificity for a biocatalyst, and stability under normal storage conditions. Sample analyses need to be minimally invasive and free from sample pre-treatment or allow minimal sample pre-treatment. The functioning of the biosensor should remain independent of physical parameters. The sensing output needs to be precise, reproducible, and linear over wide concentration ranges and should not have electrical or transducer induced noise. It should facilitate rapid, sensitive, selective, and accurate analysis of multiple target analytes simultaneously in multiplexed and cost-effective manner without the requirement of sophisticated laboratory setup and trained or skilled person.

References

Abdalla, N.S., Amr, A.E.G.E., El-Tantawy, A.S.M., Al-Omar, M.A., Kamel, A.H., Khalifa, N.M., 2019. Tailor-made specific recognition of cyromazine pesticide integrated in a potentiometric strip cell for environmental and food analysis. Polymer (Basel) 11, 1526.

Ahmadalinezhad, A., Wu, G., Chen, A., 2011. Mediator-free electrochemical biosensor based on buckypaper with enhanced stability and sensitivity for glucose detection. Biosens. Bioelectron. 30, 287–293.

Aizawa, M., Morioka, A., Suzuki, S., Nagamura, Y., 1979. Enzyme immunosensor. III. Amperometric determination of human chorionic gonadotropin by membrane-bound antibody. Anal. Biochem. 94, 22–28.

Alamer, S., Eissa, S., Chinnappan, R., Herron, P., Zourob, M., 2018. Rapid colorimetric lactoferrin-based sandwich immunoassay on cotton swabs for the detection of foodborne pathogenic bacteria. Talanta 185, 275–280.

Algar, W.R., Khachatrian, A., Melinger, J.S., Huston, A.L., Stewart, M.H., Susumu, K., Blanco-Canosa, J.B., Oh, E., Dawson, P.E., Medintz, I.L., 2017. Concurrent modulation of quantum dot photoluminescence using a combination of charge transfer and Förster resonance energy transfer: competitive quenching and multiplexed biosensing modality. J. Am. Chem. Soc. 139, 363–372.

Ali, A.A., Altemimi, A.B., Alhelfi, N., Ibrahim, S.A., 2020. Application of biosensors for detection of pathogenic food bacteria: a review. Biosensors 10, 58.

Ali, J., Najeeb, J., Ali, M.A., Aslam, M.F., Raza, A., 2017. Biosensors: their fundamentals, designs, types and most recent impactful applications: a review. J. Biosens. Bioelectron. 8, 1–9.

Alizadeh, N., Salimi, A., Hallaj, R., 2017. Hemin/G-quadruplex horseradish peroxidase-mimicking DNAzyme: principle and biosensing application. In: Advances in Biochemical Engineering/Biotechnology. Springer, Berlin, Heidelberg, pp. 1–22.

Amiri, M., Bezaatpour, A., Jafari, H., Boukherroub, R., Szunerits, S., 2018. Electrochemical methodologies for the detection of pathogens. ACS Sens. 3, 1069–1086.

Aracil, C., Perdigones, F., Moreno, J., Luque, A., Quero, J., 2015. Portable lab-on-PCB platform for autonomous micromixing. Microelectron. Eng. 131, 13–18.

Arduini, F., Cinti, S., Scognamiglio, V., Moscone, D., 2016. Nanomaterials in electrochemical biosensors for pesticide detection: advances and challenges in food analysis. Microchim. Acta 183, 2063–2083.

Arlett, J.L., Myers, E.B., Roukes, M.L., 2011. Comparative advantages of mechanical biosensors. Nat. Nanotechnol. 6, 203.

Badihi-Mossberg, M., Buchner, V., Rishpon, J., 2007. Electrochemical biosensors for pollutants in the environment. Electroanalysis 19, 2015–2028.

Bagal-Kestwal, D.R., Pan, M.H., Chiang, B.H., 2018. Electrically nanowired-enzymes for probe modification and sensor fabrication. Biosens. Bioelectron. 121, 223–235.

Bai, H., Xu, Y., Zhao, L., Li, C., Shi, G., 2009. Non-covalent functionalization of graphene sheets by sulfonated polyaniline. Chem. Commun. 13, 1667.

Bakker, E., Qin, Y., 2006. Electrochemical sensors. Anal. Chem. 78, 3965.

Bard, A.J., Faulkner, L.R., 1980. Electrochemical Methods: Fundamentals and Applications. John Wiley and Sons Inc, New York, pp. 413–414.

Barry, R.C., Lin, Y., Wang, J., Liu, G., Timchalk, C.A., 2009. Nanotechnology-based electrochemical sensors for biomonitoring chemical exposures. J. Expo Sci. Environ. Epidemiol. 19, 1–18.

Bayramoglu, G., Ozalp, V.C., Oztekin, M., Arica, M.Y., 2019. Rapid and label-free detection of *Brucella melitensis* in milk and milk products using an aptasensor. Talanta 200, 263–271.

Beasley, V.R., Coppock, R.W., Simon, J., Ely, R., Buck, W.B., Corley, R.A., Carlson, D.M., Gorham, P.R., 1983. Apparent blue-green algae poisoning in swine subsequent to ingestion of a bloom dominated by anabaena spiroides. J. Am. Vet. Med. Assoc. 182, 413–414.

Bhalla, N., Jolly, P., Formisano, N., Estrela, P., 2016. Introduction to biosensors. Essays Biochem. 60, 1–8.

Blandón-Naranjo, L., Della Pelle, F., Vázquez, M.V., Gallego, J., Santamaría, A., Alzate-Tobón, M., Compagnone, D., 2018b. Electrochemical behaviour of microwave-assisted oxidized MW-CNTs based disposable electrodes: proposal of a NADH electrochemical sensor. Electroanalysis 30, 509–516.

Blandón-Naranjo, L., Hoyos-Arbeláez, J., Vázquez, M.V., Della Pelle, F., Compagnone, D., 2018a. NADH oxidation onto different carbon-based sensors: effect of structure and surface-oxygenated groups. J. Sens. 2018, 6525919.

Blasco, A.J., González Crevillén, A., González, M.C., Escarpa, A., 2007. Direct electrochemical sensing and detection of natural antioxidants and antioxidant capacity in vitro systems. Electroanalysis 19, 2275–2286.

Bollella, P., Fusco, G., Tortolini, C., Sanzò, G., Favero, G., Gorton, L., Antiochia, R., 2017. Beyond graphene: electrochemical sensors and biosensors for biomarkers detection. Biosens. Bioelectron. 89, 152–166.

Bu, D., Zhuang, H., Yang, G., Ping, X., 2014. An immunosensor designed for polybrominated biphenyl detection based on fluorescence resonance energy transfer (FRET) between carbon dots and gold nanoparticles. Sens. Actuators B Chem. 195, 540–548.

Campuzano, S., Pedrero, M., Nikoleli, G.P., Pingarrón, J.M., Nikolelis, D.P., 2017a. Hybrid 2D-nanomaterials-based electrochemical immunosensing strategies for clinical biomarkers determination. Biosens. Bioelectron. 89, 269–279.

Campuzano, S., Pedrero, M., Pingarrón, J.M., 2017b. Non-invasive breast cancer diagnosis through electrochemical biosensing at different molecular levels. Sensors 17, 1993.

Campuzano, S., Yáñez-Sedeño, P., Pingarrón, J.M., 2017c. Electrochemical Genosensing of circulating biomarkers. Sensors 17, 866.

Campuzano, S., Yáñez-Sedeño, P., Pingarrón, J.M., 2017d. Molecular biosensors for electrochemical detection of infectious pathogens in liquid biopsies: current trends and challenges. Sensors 17, 2533.

Chen, G., Lin, Y.H., Wang, J., 2006. Monitoring environmental pollutants by microchip capillary electrophoresis with electrochemical detection. Talanta 68, 497–503.

Chen, X., Wang, Y., Zhou, J., Yan, W., Li, X., Zhu, J.J., 2008. Electrochemical impedance immunosensor based on three-dimensionally ordered macroporous gold film. Anal. Chem. 80, 2133–2140.

Cheng, B., Cunningham, B., Boeras, D., Mafaune, P., Simbi, R., Peeling, R., 2016. Data connectivity: a critical tool for external quality assessment. Afr. J. Lab. Med. 5, 535.

Chinnappan, R., Eissa, S., Alotaibi, A., Siddiqua, A., Alsager, O.A., Zourob, M., 2020. In vitro selection of DNA aptamers and their integration in a competitive voltammetric biosensor for azlocillin determination in waste water. Anal. Chim. Acta 1101, 149–156.

Chiriacò, M., Parlangeli, I., Sirsi, F., Poltronieri, P., Primiceri, E., 2018. Impedance sensing platform for detection of the food pathogen listeria monocytogenes. Electronics 7, 347.

Choi, J., Seong, T., Jeun, M., Lee, K., 2017. Field-effect biosensors for on-site detection: recent advances and promising targets. Adv. Healthc. Mater. 6, 1700796.

Cieplak, M., Kutner, W., 2016. Artificial biosensors: how can molecular imprinting mimic biorecognition? Trends Biotechnol. 34, 922–941.

Clark, L.C., Lyons, C., 1962. Electrode systems for continuous monitoring in cardiovascular surgery. Ann. N. Y. Acad. Sci. 102, 29–45.

Cunha, I., Biltes, R., Sales, M., Vasconcelos, V., 2018. Aptamer-based biosensors to detect aquatic phycotoxins and cyanotoxins. Sensors 18, 2367.

Dabek-Zlotorzynska, E., Chen, H., Ding, L.Y., 2003. Recent advances in capillary electrophoresis and capillary electrochromatography of pollutants. Electrophoresis 24, 4128–4149.

Di Gleria, K., Green, M.J., Hill, H.A.O., McNeil, C.J., 1986. Homogeneous ferrocene mediated amperometric biosensors. Anal. Chem. 58, 1203–1205.

Ding, R., Lisak, G., 2019. Sponge-based microfluidic sampling for potentiometric ion sensing. Anal. Chim. Acta 1091, 103–111.

Doria, G., Conde, J., Veigas, B., Giestas, L., Almeida, C., Assunção, M., Rosa, J., Baptista, P.V., 2012. Noble metal nanoparticles for biosensing applications. Sensors 12, 1657–1687.

Eissa, S., Ng, A., Siaj, M., Zourob, M., 2014. Label-free voltammetric aptasensor for the sensitive detection of microcystin-LR using graphene-modified electrodes. Anal. Chem. 86, 7551–7557.

El-Ansary, A., Faddah, L.M., 2010. Nanoparticles as biochemical sensors. Nanotechnol. Sci. Appl. 3, 65–76.

Escarpa, A., 2014. Lights and shadows on food microfluidics. Lab Chip 14, 3213–3224.

Fekry, A.M., Shehata, M., Azab, S.M., Walcarius, A., 2020. Voltammetric detection of caffeine in pharmacological and beverages samples based on simple nano- co (II, III) oxide modified carbon paste electrode in aqueous and micellar media. Sens. Actuators B Chem. 302, 127172.

Felix, F.S., Angnes, L., 2018. Electrochemical immunosensors - a powerful tool for analytical applications. Biosens. Bioelectron. 15 (102), 470–478.

Ferrigno, P.K., 2016. Non-antibody protein-based biosensors. Essays Biochem. 60, 19–25.

Fiel, W.A., de Freitas Borges, P.A., Lins, F.C., de Faria, R.A.D., 2019. Recent advances on the electrochemical transduction techniques for the biosensing of pharmaceuticals in aquatic environments. Int. J. Biosens. Bioelectron. 5, 119–123.

Fu, Y.Z., Yuan, R., Tang, D.P., Chai, Y.Q., Xu, L., 2005. Study on the immobilization of anti-IgG on Au-colloid modified gold electrode via potentiometric immunosensor, cyclic voltammetry, and electrochemical impedance techniques. Colloids Surf. B 40, 61–66.

Gerard, M., Chaubey, A., Malhotra, B.D., 2002. Application of conducting polymers to biosensors. Biosens. Bioelectron. 17, 345–359.

Ghindilis, L., Atanasov, P., Wilkinst, M., Wilkins, E., 1998. Immunosensors: electrochemical sensing and other engineering approaches. Biosens. Bioelectron. 13, 113.

Ghindilis, A.L., Skorobotko, O.V., Gavnlova, V.P., Yaropolov, A.I., 1992. A new approach to the construction of potentiometric immunosensors. Biosens. Bioelectron. 7, 301–304.

Gilbert, D.M., Sale, T.C., 2005. Sequential electrolytic oxidation and reduction of aqueous phase energetic compounds. Environ. Sci. Technol. 39, 9270–9277.

Guilbault, G.G., Kramer, D.N., Cannon, P.L., 1962. Electrochemical determination of organophosphorus compounds. Anal. Chem. 34, 1437–1439.

Guilbault, G.G., Montalvo Jr., J.G., 1969. Urea specific enzyme electrode. J. Am. Chem. Soc. 91, 2164–2168.

Hameed, S., Xie, L., Ying, Y., 2018. Conventional and emerging detection techniques for pathogenic bacteria in food science: a review. Trends Food Sci. Technol. 81, 61–73.

Hayat, A., Marty, J.L., 2014. Aptamer based electrochemical sensors for emerging environmental pollutants. Front. Chem. 2, 41.

He, Z., Wei, J., Gan, C., Liu, W., Liu, Y., 2017. A rolling circle amplification signal-enhanced immunosensor for ultrasensitive microcystin-LR detection based on a magnetic graphene functionalized electrode. RSC Adv. 7, 39906–39913.

Heineman, W.R., Kissinger, P.T., Kissinger, P.T., 2018. Large-Amplitude Controlled-Current Techniques. In: Laboratory Techniques in Electroanalytical Chemistry, Revised and Expanded. CRC Press, pp. 127–140.

Himmelstoß, S.F., Wiesholler, L.M., Buchner, M., Muhr, V., Märkl, S., Baeumner, A.J., Hirsch, T., 2017. 980 nm and 808 nm excitable upconversion nanoparticles for the detection of enzyme related reactions. Proc. SPIE, 10077.

Holzinger, M., Goff, A.L., Cosnier, S., 2017. Synergetic effects of combined nanomaterials for biosensing applications. Sensors 17, 1010.

Hu, T., Mei, X., Wang, Y., Weng, X., Liang, R., Wei, M., 2019. Two-dimensional nanomaterials: fascinating materials in biomedical field. Sci. Bull. 64, 1707–1727.

Jiang, H., Jiang, D., Zhu, P., Pi, F., Ji, J., Sun, C., Sun, J., Sun, X., 2016. A novel mast cell co-culture microfluidic chip for the electrochemical evaluation of food allergen. Biosens. Bioelectron. 83, 126–133.

Jolly, P., Formisano, N., Tká˘c, J., Kasák, P., Frost, C., Estrela, P., 2015. Label-free impedimetric aptasensor with antifouling surface chemistry: a prostate specific antigen case study. Sens. Actuators B Chem. 209, 306–312.

Jolly, P., Rainbow, J., Regoutz, A., Estrela, P., Moschou, D., 2019. A PNA-based lab-on-PCB diagnostic platform for rapid and high sensitivity DNA quantification. Biosens. Bioelectron. 123, 244–250.

Jothimuthu, P., Wilson, R.A., Herren, J., Haynes, E.N., Heineman, W.R., Papautsky, I., 2011. Lab-on-a-chip sensor for detection of highly electronegative heavy metals by anodic stripping voltammetry. Biomed. Microdevices 13, 695–703.

Kharitonov, A.B., Alfonta, L., Katz, E., Willner, I., 2000. Probing of bioaffinity interactions at interfaces using impedance spectroscopy and chronopotentiometry. J. Electroanal. Chem. 487, 133–141.

Kissinger, P.T., Ridgway, T.H., 2019. Small-amplitude controlled-potential techniques. In: Laboratory Techniques in Electroanalytical Chemistry, Revised and Expanded. CRC Press, pp. 141–164.

Land, K., Boeras, D., Chen, X., Ramsay, A., Peeling, R., 2018. REASSURED diagnostics to inform disease control strategies, strengthen health systems and improve patient outcomes. Nat. Microbiol. 4, 46–54.

Lee, M., Lee, K., Kim, K., Oh, K., Choo, J., 2012. SERS-based immunoassay using a gold array-embedded gradient microfluidic chip. Lab Chip 12, 3720.

Lin, J., Ju, H., 2005. Electrochemical and chemiluminescent immunosensors for tumor markers. Biosens. Bioelectron. 20, 1461.

Liu, G.D., Lin, Y.H., 2005. Electrochemical sensor for organophosphate pesticides and nerve agents using zirconia nanoparticles as selective sorbents. Anal. Chem. 77, 5894–5901.

Liu, G., Lin, Y., 2007. Nanomaterial labels in electrochemical immunosensors and immunoassays. Talanta 74, 308.

Liu, G., Wang, J., Wu, H., Lin, Y., Lin, Y., 2007. Nanovehicles based bioassay labels. Electroanalysis 19, 777.

Liu, B.Q., Zhang, B., Cui, Y.L., 2011. Multifunctional gold-silica nanostructures for ultrasensitive electrochemical immunoassay of streptomycin residues. ACS Appl. Mater. Interfaces 3, 4668–4676.

López, M.A., Moreno-Guzman, M., Jurado, B., Escarpa, A., 2016. Food microfluidics biosensors. In: Comprehensive Analytical Chemistry. Elsevier, Amsterdam, The Netherlands.

Ma, X., Chen, M., 2015. Electrochemical sensor based on graphene doped gold nanoparticles modified electrode for detection of diethylstilboestrol. Sensors Actuators B Chem. 215, 445–450.

MacKintosh, C., Beattie, K.A., Klumpp, S., Cohen, P., Codd, G.A., 1990. Cyanobacterial microcystin-LR is a potent and specific inhibitor of protein phosphatases 1 and 2A from both mammals and higher plants. FEBS Lett. 264, 187–192.

Maduraiveeran, G., Sasidharan, M., Ganesan, V., 2018. Electrochemical sensor and biosensor platforms based on advanced nanomaterials for biological and biomedical applications. Biosens. Bioelectron. 103, 113–129.

Manfredi, A., Giannetto, M., Mattarozzi, M., Costantini, M., Mucchino, C., Careri, M., 2016. Competitive immunosensor based on gliadin immobilization on disposable carbon-nanogold screen-printed electrodes for rapid determination of celiotoxic prolamins. Anal. Bioanal. Chem. 408, 7289–7298.

Marquette, C.A., Blum, L.J., 2006. State of the art and recent advances in immunoanalytical systems. Biosens. Bioelectron. 21, 1424.

Martynenko, I.V., Litvin, A.P., Purcell-Milton, F., Baranov, A.V., Fedorov, A.V., Gun'ko, Y.K., 2017. Application of semiconductor quantum dots in bioimaging and biosensing. J. Mater. Chem. B 5, 6701–6727.

Mayedwa, N., Ajayi, R.F., Mongwaketsi, N., Matinise, N., Mulaudzi-Masuku, T., Hendricks, K., Maaza, M., 2019. Development of a novel tyrosinase amperometric biosensor based on tin nanoparticles for the detection of bisphenol A (4.4-isopropylidenediphenol) in water. J. Phys. Conf. Ser. 1310, 12005.

Meisam, O., Malakoutian, M.A., Mohammadmehdi, C., Oroojalian, F., Haghiralsadat, F., Yazdian, F., 2013. A label-free detection of biomolecules using micromechanical biosensors. Chin. Phys. Lett. 30, 068701.

Mirsky, V.M., Riepl, M., Wolfbeis, O.S., 1997. Capacitive monitoring of protein immobilization and antigen-antibody reactions on monomolecularalkylthiol films on gold electrodes. Biosens. Bioelectron. 12, 977–989.

Moro, L., Turemis, M., Marini, B., Ippodrino, R., Giardi, M.T., 2017. Better together: strategies based on magnetic particles and quantum dots for improved biosensing. Biotechnol. Adv. 35, 51–63.

Moschou, D., Tserepi, A., 2017. The lab-on-PCB approach: tackling the μTAS commercial upscaling bottleneck. Lab Chip 17, 1388–1405.

Muhammad-Tahir, Z., Alocilja, E.C., 2003. A conductometric biosensor for biosecurity. Biosens. Bioelectron. 18, 813–819.

Mustafa, F., Hassan, R.Y.A., Andreescu, S., 2017. Multifunctional nanotechnology-enabled sensors for rapid capture and detection of pathogens. Sensors 17, 2121.

Neethirajan, S., Raghavan, V., Weng, X., Chand, R., 2018. Biosensors for sustainable food engineering: challenges and perspectives. Biosensors 8, 23.

Neufeld, T., Eshkenazi, I., Cohen, E., Rishpon, J., 2000. A micro flow injection electrochemical biosensor for organophosphorus pesticides. Biosens. Bioelectron. 15, 323–329.

Oliveira Brett, A.M., Diculescu, V.C., Chiorcea-Paquim, A.M., 2007. Serrano SH chapter 20 DNAElectrochemical biosensors for investigating DNA damage. Compr. Anal. Chem. 49, 413–437.

Ou, X., Chen, P., Huang, X., Li, S., Liu, B., 2019. Microfluidic chip electrophoresis for biochemical analysis. J. Sep. Sci. 43, 258–270.

Owen, D., Mao, W., Alexeev, A., Cannon, J., Hesketh, P., 2013. Microbeads for sampling and mixing in a complex sample. Micromachines 4, 103–115.

Pan, M., Yagauang, Y.G., Li, M., Zing, X., Wang, S., 2017. Nanomaterials for electrochemical immunosensing. Sensors 17, 1041.

Qiu, H.J., Guan, Y., Luo, P., Wang, Y., 2017. Recent advance in fabricating monolithic 3D porous graphene and their applications in biosensing and biofuel cells. Biosens. Bioelectron. 89, 85–95.

Rainbow, J., Sedlackova, E., Jiang, S., Maxted, G., Moschou, D., Richtera, L., Estrela, P., 2020. Integrated electrochemical biosensors for detection of waterborne pathogens in low-resource settings. Biosensors 10, 36.

Rapini, R., Marrazza, G., 2017. Electrochemical aptasensors for contaminants detection in food and environment: recent advances. Bioelectrochemistry 118, 47–61.

Rasheed, P.A., Sandhyarani, N., 2017. Electrochemical DNA sensors based on the use of gold nanoparticles: a review on recent developments. Microchim. Acta 184, 981–1000.

Renedo, O.D., Alonso-Lomillo, M.A., Martinez, M.J.A., 2007. Recent developments in the field of screen-printed electrodes and their related applications. Talanta 73, 202–219.

Rogers, K.R., 2000. Principles of affinity-based biosensors. Mol. Biotechnol. 14, 109–130.

Ronkainen, N.J., Halsall, H.B., Heineman, W.R., 2010. Electrochemical biosensors. Chem. Soc. Rev. 39, 1747–1763.

Sankarankutty, K.M., 2014. Biosensors and their applications for ensuring food safety. Glob. J Pathol Microbiol. 2, 15–21.

Santos, P.V.F., Lopes, I.C., Diculescu, V.C., Oliveira-Brett, A.M., 2012. DNA—cyanobacterial hepatotoxins microcystin-LR and nodularin interaction: electrochemical evaluation. Electroanalysis 24, 547–553.

Sassolas, A., Blum, L.J., Leca-Bouvier, B.D., 2012. Immobilization strategies to develop enzymatic biosensors. Biotechnol. Adv. 30, 489–511.

Shanko, E., van de Burgt, Y., Anderson, P., den Toonder, J., 2019. Microfluidic magnetic mixing at low Reynolds numbers and in stagnant fluids. Micromachines 10, 731.

Sharma, S., Byrne, H., O'Kennedy, R.J., 2016. Antibodies and antibody-derived analytical biosensors. Essays Biochem. 60, 9–18.

Shi, H., Wei, J., Qiang, L., Chen, X., Meng, X., 2014. Fluorescent carbon dots for biolmaging and biosensing applications. J. Biomed. Nanotechnol. 10, 2677–2699.

Singh, P., 2014. Surface Plasmon Resonance. Nova Science Publishers, New York, USA.

Singh, P., 2016a. Chemical, SPR biosensors: historical perspectives and current challenges. Sens. Actuators B 229, 110–130.

Singh, P., 2016b. LSPR biosensing: recent advances and approaches in reviews. In: Giddes, C.D. (Ed.), Plasmonics. vol. 2017. Springer International Publishing AG, pp. 211–238.

Singh, P., 2017a. G protein coupled receptors: a potential candidate for drug discovery. Intl. J. Pharmacol. 13 (Special Issue), 846–863.

Singh, P., 2017b. Surface Plasmon Resonance: A Boon for Viral Diagnostics in Reference Module in Life Sciences. Elsevier, Netherlands.

Skoog, D.A., Leary, J.J., 1992. Principles of Instrumental Analysis, fourth ed. Harcourt Brace College Publishers, New York, pp. 535–564.

Smith, J.P., Foster, C.W., Metters, J.P., Sutcliffe, O.B., Banks, C.E., 2014. Metallic impurities in graphene screen-printed electrodes can influence their electrochemical properties. Electroanalysis 26, 2429–2433.

Sole, S., Alegret, S., 2001. Environmental toxicity monitoring using electrochemical biosensing systems. Environ. Sci. Pollut. Res. Int. 8, 256–264.

Stradiotto, N.R., Yamanaka, H., Zanoni, M.V.B., 2003. Electrochemical sensors: a powerful tool in analytical chemistry. J. Braz. Chem. Soc. 14, 159.

Tang, J., Huang, J.X., Su, B.L., Chen, H.F., Tang, D.P., 2011. Sandwich-type conductometric immunoassay of alpha-fetoprotein in human serum using carbon nanoparticles as labels. Biochem. Eng. J. 53, 223–228.

Tang, D.P., Yuan, R., Chai, Y.Q., Liu, Y., Zhong, X., Dai, J.Y., 2004. Potentiometric diphtheria immunosensor based on a glassy carbon electrode modified with colloidal gold and anti-diph. Acta Chim. Sin. 62, 2062–2066.

Updike, S.J., Hicks, G.P., 1967. The enzyme electrode. Nature 214, 986–988.

Vogiazi, V., de la Cruz, A., Mishra, S., Shanovc, V., Heinemanb, W.R., Dionysiou, D.D., 2019. A comprehensive review: development of electrochemical biosensors for detection of cyanotoxins in freshwater. ACS Sens. 4 (5), 1151–1173.

Wanekaya, A.K., 2011. Applications of nanoscale carbon-based materials in heavy metal sensing and detection. Analyst 136, 4383–4391.

Wang, J., 2007. Nanopraticle based electrochemical bioassays of proteins. Electroanalysis 19, 769.

Wang, J., 2018. Electrochemical preconcentration. In: Laboratory Techniques in Electroanalytical Chemistry, Revised and Expanded. CRC Press, pp. 719–738.

Wang, J., Chatrathi, M.P., Mulchandani, A., Chen, W., 2001. Capillary electrophoresis microchips for separation and detection of organophosphate nerve agents. Anal. Chem. 73, 1804–1808.

Wang, J., Liu, G.D., Jan, M.R., 2004. Ultrasensitive electrical biosensing of proteins and DNA: carbon-nanotube derived amplification of the recognition and transduction events. J. Am. Chem. Soc. 126, 3010–3011.

Wang, Y., Partridge, A., Wu, Y., 2019. Improving nanoparticle-enhanced surface plasmon resonance detection of small molecules by reducing steric hindrance via molecular linkers. Talanta 1 (198), 350–357.

Wang, J., Pumera, M., Chatrathi, M.P., Escarpa, A., Musameh, M., Collins, G., Mulchandani, A., Lin, Y., Olsen, K., 2002. Single-channel microchip for fast screening and detailed identification of nitroaromatic explosives or organophosphate nerve agents. Anal. Chem. 74, 1187–1191.

Wang, Y., Tang, L.J., Jiang, J.H., 2013a. Surface-enhanced Raman spectroscopy-based, homogeneous, multiplexed immunoassay with antibody-fragments-decorated gold nanoparticles. Anal. Chem. 85, 9213–9220.

Wang, T., Zhu, H., Zhuo, J., Zhu, Z., Papakonstantinou, P., Lubarsky, G., Lin, J., Meixian Li, M., 2013b. Biosensor based on ultrasmall MoS2 nanoparticles for electrochemical detection of H_2O_2 released by cells at the nanomolar level. Anal. Chem. 85, 10289–10295.

Warsinke, A., Benkert, A., Scheller, F.W., 2000. Electrochemical immunoassays. J. Fresenius Anal. Chem. 366, 622.

Wu, Q., Zhang, Y., Yang, Q., Yuan, N., Zhang, W., 2019. Review of electrochemical DNA biosensors for detecting food borne pathogens. Sensors (Basel) 19 (22), 4916.

Xiong, M., Rong, Q., Meng, H.M., Zhang, X.B., 2017. Two-dimensional graphitic carbon nitride nanosheets for biosensing applications. Biosens. Bioelectron. 89, 212–223.

Xu, T.S., 2016. Bioconjugation of peroxidase-like nanostructures with natural enzyme for in-situ amplified conductometric immunoassay of tissue polypeptide antigen in biological fluids. Biochem. Eng. J. 105, 36–43.

Xu, J.J., Wang, A.J., Chen, H.Y., 2007. Electrochemical detection modes for microchip capillary electrophoresis. TrAC Trends Anal. Chem. 26, 125–132.

Yagiuda, K., Hemmi, A., Ito, S., Asano, Y., 1996. Development of a conductivity-based immunosensor for sensitive detection of methamphetamine (stimulant drug) in human urine. Biosens. Bioelectron. 11, 703–707.

Yáñez-Sedeño, P., Campuzano, S., Pingarrón, J.M., 2019. Pushing the limits of electrochemistry toward challenging applications in clinical diagnosis, prognosis, and therapeutic action. Chem. Commun. (Camb) 55 (18), 2563–2592. 26.

Yu, X., Chen, F., Wang, R., Li, Y., 2018. Whole-bacterium SELEX of DNA aptamers for rapid detection of *E. coli* O157: H7 using a QCM sensor. J. Biotechnol. 266, 39–49.

Yu, X., Munge, B., Patel, V., Jensen, G., Bhirde, A., Gong, J.D., Kim, S.N., Gillespie, J., Gutkind, J.S., Papadimitrakopoulos, F., Rusling, J.F., 2006. Carbon nanotube amplification strategies for highly sensitive immunodetection of cancer biomarkers. J. Am. Chem. Soc. 128, 11199–11205.

Yu, S., Xu, J., Huang, K., Chen, J., Duan, J., Xu, Y., Qing, H., Geng, L., Deng, Y., 2016. A novel method to predict protein aggregations using two-dimensional native protein microfluidic chip electrophoresis. Anal. Methods 8, 8306–8313.

Yu, X., Zhang, W., Zhang, P., Su, Z., 2017. Fabrication technologies and sensing applications of graphene-based composite films: advances and challenges. Biosens. Bioelectron. 89, 72–84.

Zhang, H., Chang, H., Neuzil, P., 2019. DEP-on-a-chip: dielectrophoresis applied to microfluidic platforms. Micromachines 10, 423.

Zhang, Y., Chen, M., Li, H., Yan, F., Pang, P., Wang, H., Wu, Z., Yang, W., 2017. A molybdenum disulfide/gold nanorod composite-based electrochemical immunosensor for sensitive and quantitative detection of microcystin-LR in environmental samples. Sens. Actuators B 244, 606–615.

Zhang, J., Sun, Y., Dong, H., Zhang, X., Wang, W., Chen, Z., 2016. An electrochemical nonenzymatic immunosensor for ultrasensitive detection of microcystin-LR using carbon nanofibers as the matrix. Sens. Actuators B 233, 624–632.

Zhao, X., Zhuang, H., Yoon, S.-C., Dong, Y., Wang, W., Zhao, W., 2017. Electrical impedance spectroscopy for quality assessment of meat and fish: a review on basic principles, measurement methods, and recent advances. J. Food Qual. 2017, 6370739.

Zhu, N., Chang, Z., He, P., Fang, Y., 2005. Electrochemical DNA biosensors based on platinum NPs combined carbon nanotubes. Anal. Chim. Acta 545, 21–26.

Zhu, Z., Zhang, Y.H.P., 2015. Use of nonimmobilized enzymes and mediators achieved high power densities in closed biobatteries. Energy Sci. Eng. 3, 490–497.

Zhuo, Z., Ma, M.N., Chai, Y.Q., Zhao, M., Yuan, R., 2014. Amplified electrochemiluminescent aptasensor using mimicking bi-enzyme nanocomplexes as signal enhancement. Anal. Chim. Acta 809, 47–53.

Zhuo, Y., Yuan, R., Chai, Y.Q., Hong, C.L., 2010. Functionalized SiO2 labeled CA19–9 antibodies: a new strategy for signal amplification of antigen-antibody sensing processes. Analyst 135, 2036–2042.

Chapter 2

Nanomaterial based electrochemical biosensing: Progress and perspectives

2.1. Introduction

Rapid and progressive development of nanotechnology has ensured the design and synthesis of novel electrochemical biosensors with improvised features (Campuzano et al., 2017c). Recent developments in the field of molecular biology, nanofabrication protocols, labeling, nano-instrumentation, multiplexing capabilities, along with microfluidics has led to the development of easy-to-use electrochemical sensors enabling rapid, sensitive, selective, detection of circulating biomarkers in the clinical field, adulterants in food, and environmental monitoring (Campuzano et al., 2017d).

This chapter critically examines the different types of nanomaterials and their role in different biosensing platforms such as optical, electrochemical, mechanical, etc. in entirety. Various types of nanomaterials are discussed in their exhaustive details that include their synthesis, and their role in a variety of biosensing strategies for their use in the determination of clinically relevant biomarkers, adulterants, and additives in food, and environmental biomonitoring. The chapter also analyzes the challenges that remain in terms of ease of fabrication of NM-based biosensors, its integration with other biosensing platforms, and its universal applicability in resource-limited set-ups in remote areas. Relevant work of literature spanning over the period of last 20 years are incorporated to present the work in its exhaustive details.

The major aspects of the entire chapter incorporate the following: (i) electrochemical biosensors, its major types, (ii) use of different nanoparticles (NPs) for fabrication and function of various biosensing platforms, (iii) use of different nanomaterials (NMs) for fabrication and function of various biosensing platforms, (iv) use of carbon-based nanomaterials for fabrication and function of various biosensing platforms, (vi) use of semiconductor nanomaterials for fabrication and function of various biosensing platforms, and (vii) use of other nanomaterials for fabrication and function of various biosensing platforms.

Electrochemical Biosensors. https://doi.org/10.1016/B978-0-323-90632-6.00005-4

2.1.1 Electrochemical immunosensors

Electrochemical immunosensors integrates traditional immunoassay with modern biosensors and electrochemical analysis, thus incorporating high selectivity of immunoassays and high sensitivity of the electrochemical analysis. The electrochemical signals can be electric current, potential, conductance, and impedance. The working of immunosensors depends on the specific recognition event between antigen and antibody. The biological molecules such as proteins, hormones, antibiotics, persistent environmental pollutants, and toxins act as haptens triggering the release of specific antibodies ensuring high selectivity of the reaction. Immunosensors can be of electrochemical, optical, piezoelectric, or calorimetric types, etc. (Pan et al., 2017; Singh, 2014, 2016a, b; Singh, 2017a, b). Electrochemical biosensors offer affordable, reagent-less monitoring, and identification of biomarkers at different molecular levels in samples of different natures in a real-time manner. This means that the procedure can be carried out even in resource-limited remote settings. These are advantageous, particularly for complex clinical and environmental samples, such as undiluted liquid biopsies, whole cells, and paraffin-embedded tissues, water samples, which are difficult to analyze using conventional techniques, and also issues arising due to matrix effects (Pan et al., 2017; Singh, 2014, 2016a, b; Singh, 2017a, b).

2.1.2 Type of electrochemical immunosensors based on signal

Electrochemical immunosensors analyzes electrical signal (voltammetric, potentiometric, conductometric or impedimetric) produced on an electrochemical transductor. Aizawa et al. (1979) have pioneered the concept of an **Amperometric immunosensor**. This type of sensor measures the changes in current, which is directly proportional to the analyte concentration at constant voltage indicating the redox reaction on the analyte. Amperometric immunosensing platforms can be of two types, sandwich-type, and competitive type. The concentration of electroactive materials like, enzymes, antigens, or antibodies can be determined by labeling and monitoring the current change. The current signal can be enhanced by the markers such as alkaline phosphatase (ALP), horseradish peroxidase (HRP) or glucose oxidase, or nanomaterials (Au nanoparticles (AuNPs), graphene, carbon nanotubes (CNTs), or silica nanoparticles. **Potentiometric Immunosensors** were first reported by Ghindilis et al. (1992) for the detection of insulin. Potentiometric immunosensors perform real-time monitoring of potential change for the direct or indirect determination of antigen. There is a logarithmic relationship between the potential change and antigen or antibody concentration. Potentiometric immunosensors usually measure the membrane or ion-electrode potential. (Kharitonov et al., 2000; Fu et al., 2005; Tang et al., 2004). **Impedance Immunosensors** measure the label-free, real-time change in the transfer of electrons in solutions before and after an immune reaction to determine the test-analyte (Chen et al., 2008). These immunosensors displays the advantages of electrochemical impedance spectroscopy (EIS) such as capacitance, inductance, and resistance to produce specific impedance signal that is utilized to study the properties of conductive materials and interfaces (Zhuo et al., 2010; Yagiuda et al., 1996). **Conductometric Immunosensors** detect an antigen by monitoring the changes in the number and type of ions in the solution before and after immune reaction to detect an antigen. The quantification is achieved by immobilizing a specific enzyme on the surface of the metal electrode to determine the conductivity under the action of an electric field. (Muhammad-Tahir and Alocilja, 2003; Tang et al., 2011a, b; Xu, 2016). Mirsky et al. (1997) have pioneered the concept of **Capacitive Immunosensor**. It measures the changes in the capacitance of the interface between an electrode and testing solution. **Gravimetric sensors** are the piezoelectric devices such as quartz crystal microbalance (QCM) and micro- (or nano-) mechanical cantilever setups that detect a biological target relying on the change in the resonance frequency proportional to the mass of the analyte. It offers highly sensitive, and label-free detection of target analytes (El-Ansary and Faddah, 2010).

2.1.3 Electrochemical immunoassays

Electrochemical immunoassays have consistently shown their potential across scientific fraternity and industries for the POC diagnostics, biomonitoring of clinical biomarkers, environmental and food contaminants owing to high sensitivity, low cost, and inherent portability (Bakker and Qin, 2006; Ghindilis et al., 1998; Lin and Ju, 2005; Maduraiveeran et al., 2018; Marquette and Blum, 2006; Stradiotto et al., 2003; Warsinke et al., 2000). Electrochemical immunosensor relies on the principle of highly specific antibody-antigen interaction that involves immobilizing the immunologic material on an electrochemical transducer. An immunoassay works by immobilizing the immunologic material on a solid support such as microplate wells, magnetic beads, and polystyrene beads. Nanomaterials (NMs) are attached to the transducer surface or solid support following immunoreactions. Transducing signals of antibody-antigen reaction can be amplified using labels like enzymes, electroactive molecules, redox complexes, and metal ions. Quantification involves measuring the specific activity of an NMs after the release step or after adding a substrate. (Liu and Lin, 2007; Singh, 2017a, b; Yáñez-Sedeño et al., 2019).

2.1.4 Nanomaterial based electrochemical biosensors

2D-nanomaterials (NMs) and 2D-nanohybrids enhance the conductivity and biocompatibility of electrochemical biosensors. Clinical biomarkers based on different types of hybrid 2D-NMs are available, especially hybrid nanostructures involving graphene and other graphene-like 2D-layered nanomaterials (GLNMs). Different strategies have been optimized for utilizing 2D-nanomaterial hybrids/heterostructures for developing immunosensors for tagging or modifying electrode transducers called hybrid nanocomposites (Campuzano et al., 2017c, b).

Electrochemical immunosensors have certain limitations in terms of instability, volatility, nonreproducibility, and other shortcomings related to the identification of biomolecules like antibodies (Pan et al., 2017).

Novel functional nanomaterials (NMs) and analytical technologies have greatly advanced the electrochemical sensor and biosensing platforms/devices finding important applications in the biological, biomedical, biotechnological, clinical, environmental, and food industries (Figs. 1 and 2). NMs exhibit unique surface properties, small size, and macroscopic quantum tunneling along with unique mechanical, electrical, optical, magnetic, and catalytic properties. Structural properties like strong adsorption capacity, reliable orientation, biocompatibility, and structural compatibility enable these novel NMs in the immobilization of biomolecules and play a crucial role in the development of immunosensors (Pan et al., 2017; Maduraiveeran et al., 2018; Singh, 2016a, b).

The wide array of NMs along with surface modifications and microfabrication enables in vivo and in vitro bioanalysis through specifically planned electrode/solution interface. NM-equipped sensors offer the advantage of low cost, portability, energy efficiency, ease of fabrication, online monitoring, and simultaneous sensing capability in addition to high sensitivity, selectivity, and durability. Expertise from the consortium of diverse fields of chemistry, biology, material science, and medicine will provide the main thrust for the rapid growth, advancement, and continued improvisation in electrochemical sensing

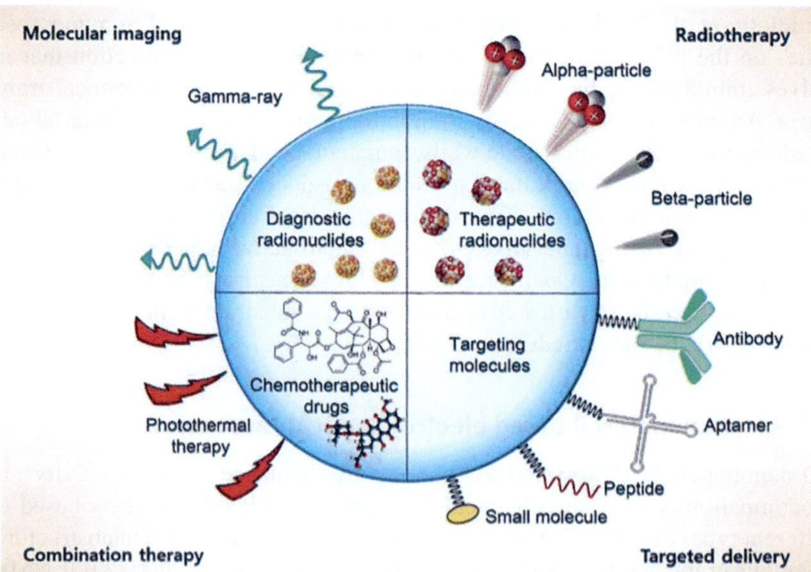

FIG. 1 Multifunctional nanomaterials for therapeutic study (Jeon, 2019). This is an open-access article distributed under the Creative Commons Attribution License which permits unrestricted use, distribution, and reproduction in any medium provided the original work is properly cited.

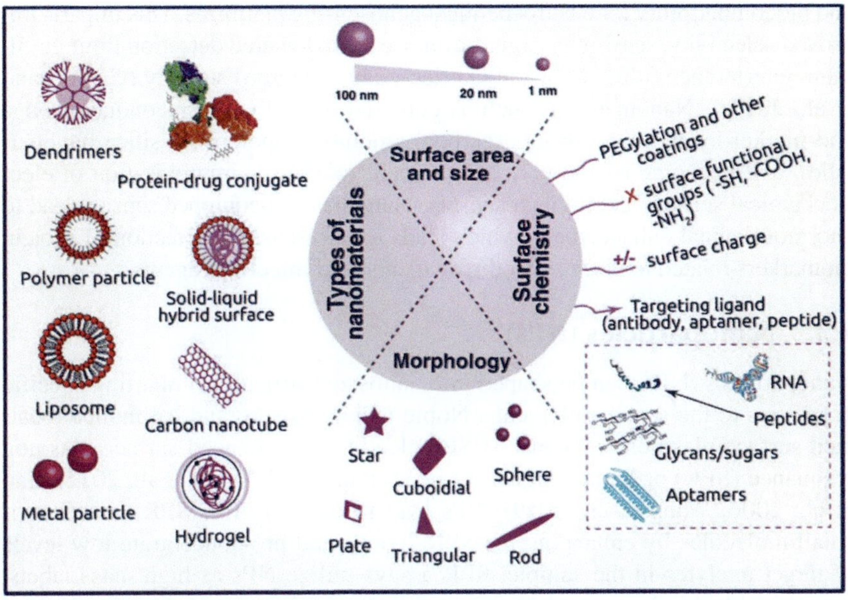

FIG. 2 Schematic representation of different types of nanomaterials employed in cancer therapy, their important physical properties, and surface chemistry required to carry drugs (Navya et al., 2019).

platforms (Maduraiveeran et al., 2018; Singh, 2016a, b). The advent of novel NMs based electrochemical markers will continue to improve the design of immunosensor (Pan et al., 2017). Nanomaterials are mainly utilized as NM-modified electrochemical transducers, and NM-bimolecular conjugates as labels in electrochemical immunosensing (Liu et al., 2007; Wang, 2007).

2.2. Types of nanomaterials

Nanomaterials such as noble metal and metal oxide nanoparticles (NPs), nanochains (NCs), nanospheres; integrated nanostructures including graphene or reduced graphene oxide (rGO), or metallic NPs or oxides, multiwalled carbon nanotubes (MWCNTs), screen-printed electrodes (SPEs), magnetic beads (MBs) exhibit unique features. Biosensors incorporating these nanomaterials are capable of detecting biomarkers with high sensitivity, accuracy, and with a lower limit of detection (LODs) (Campuzano et al., 2017c). Nanomaterials (graphene, rGO, AuNPs, AuNRs, MWCNTs) incorporated nucleic acid impart large surface area with abundant binding points, catalytic activity, conductivity, and biocompatibility for electrochemical genosensing platforms. This imparts improved selectivity, sensitivity, signal-to-noise ratio, lowered detection limit, minimum interference from biological matrices, and prolonged stability (Campuzano et al., 2017c). Nanomaterials such as colloidal gold/silver semiconductor NPs, and marker-loaded nanocarriers (carbon nanotubes, apoferritin, silica nanoparticles, and liposome beads) play a significant role in the improvisation of electrochemical sensors and immunosensors. Nanomaterial-equipped sensors lead to enormous signal enhancement, which leads to ultrasensitive detection of protein biomarkers related to disease, biothreat agents, and infectious agents.

2.3. Nanoparticles (NPs)

Nanoparticles (NPs) can be shaped into nanosized structures imparting specific properties to the sensing elements. Noble metals like Au and Ag induce localized surface plasmon resonance (LSPR) leading to enhanced surface plasmon resonance (SPR) or Raman signals (Anker et al., 2008; Doria et al., 2012; Liao et al., 2006; Wang et al., 2019). NPs lead to ultrasensitive SPR detection of small molecules by enhancing the SPR signals and pre-concentrate low levels of target analytes in the sample. SPR assays utilize NPs as high mass labels, enhancing the binding signals and detection sensitivity governed by the size and distance of the NPs from the sensing surface. Signal enhancement is directly proportional to the size of the NPs, and is inversely related to the distance of NPs from the sensing surface. Application of large NPs (≥ 100 nm) for SPR detection poses a problem as the short effective penetration depth of the SPR evanescent field, and steric hindrance of binding when small molecules are immobilized onto the surface. To address the issues of steric hindrance, Wang et al. (2019) have employed molecular linkers to tether both the antibodies to the magnetic NPs, and to bind the small molecules to the surface of the SPR chips. This extends the distance of the small molecule (progesterone) from the SPR chip surface and improv the antibody orientation on the large magnetic NPs resulting in a sensitive SPR detection for progesterone. These findings could have implications in monitoring the onset of estrus in cows. Magnetic nanoparticles (MNPs) are increasingly used in SPR bioassay. However, their extremely small size limits their detection sensitivity through conventional techniques.

2.3.1 SPR assays based on nanoparticles (NPs)

Yuan et al. (2007) have employed mixed self-assembled monolayer (mSAM) surface to bring the AuNPs close to the sensing surface in SPR assay. This group had conjugated progesterone (P4) to ovalbumin (OVA) with oligoethylene glycol (OEG) linker to form protein conjugate (P(4)-OEG-OVA), which was immobilized onto the mSAM surface. The results have shown increased detection sensitivity of progesterone with the lower limit of detection (LOD). The same mSAM/P4-OEG-OVA surface was utilized for over 400 binding/regeneration cycles without any significant drop in antibody binding capacity and baseline shift indicating the high stability of the sensor surface.

2.3.2 Nanoparticle-enhanced SPR-phase imaging (SPR-PI)

The SPR-phase imaging (SPR-PI) and nanoparticle-enhanced SPR-PI can be utilized for the bioaffinity detection of proteins and nucleic acids in a multiplexed manner. The SPR-PI experiments were performed at a near-infrared (IR) 860 nm light-emitting diode (LED) light source and a wedge depolarizer to create a phase grating on a four-element single-stranded DNA (ssDNA) microarray. Multiplexed real-time phase-shift measurements were performed to detect bioaffinity adsorption onto the various microarray elements (Mohammadzadeh-Asl et al., 2018; Yuan et al., 2007). (Zhou et al., 2012) have detected thrombin at picomolar concentrations using ssDNA aptamers via SPR-PI with 20-fold increased detection sensitivity. The group had also used SPR-PI to detect two short ssDNA oligonucleotides (38mers) simultaneously at femtomolar concentration in a three-sequence hybridization format employing 120 nm DNA-modified silica NPs to enhance the SPR-PI signal. The signal enhancement and high detection sensitivity were largely due to the adsorbed silica NPs, which greatly increases the phase shift on bioaffinity adsorption due to a large increase in the interfacial refractive index from adsorbed NPs (Zhou et al., 2012).

2.3.3 Magnetic nanoparticles (MNPs) based SPR assays

Wang et al. (2010) proposed an enhanced SPR bioassay using magnetic nanoparticles (MNPs). Although MNPs are frequently used in bio-separation, their utility in bioassays is limited due to their extremely small size, which limits the detection sensitivity using conventional techniques. The study has utilized the carboxyl group modified Fe(3)O(4) MNPs of dual sizes on SPR spectroscopy by assembling MNPs on amino group modified SPR gold substrate. SPR sensor based on an indirect competitive inhibition assay (ICIA) is developed for detecting adenosine by employing Fe(3)O(4) MNP-anti-adenosine aptamer conjugates as the amplification reagent. The results validated the role of Fe(3)O(4) MNPs as a powerful amplification agent to provide a sensitive approach to detect adenosine by SPR. The detection protocol can be easily customized to detect biomolecules of interest by changing the corresponding aptamer in Fe(3)O(4) MNP-aptamer conjugates. Equipped with this novel technique, SPR spectroscopy can be used for highly sensitive detection of MNPs based separation products and utilizing Fe(3)O(4) MNPs as an important amplification agent to fabricate high-performance SPR biosensors (Singh, 2016a, b).

2.3.4 Advantages of metal nanoparticles in SPR assays

Measuring small molecules in extremely dilute concentrations of analyte by SPR gives low detection sensitivity of small molecules. Different nanomaterials like metal nanoparticles (for instance gold, silver, and magnetic nanoparticles), quantum dots, nanorods, and carbon-based nanostructures are used to enhance SPR signals as these are easy to fabricate and come in an appropriate size and use. Additionally, SPR sensitivity could be enhanced by utilizing phase information of the reflected light, which enhances the signal. Phase-sensitive SPR sensors are advantageous and can detect infinitesimal changes in external properties of the target while traditional types of SPR cannot show these changes. This has potential in POC applications. The SPR offers a label-free sensing platform to analyze binding kinetics of bimolecular interaction in a real-time manner. These are usually based on single to several channel configurations. However, the advent of drug screening protocols coupled with the investigation of bimolecular interactions on a massive scale demands high-throughput 2D-SPR arrays that are based on imaging (SPRi). These sensing-cum-imaging platforms have important applications in molecular sensing, POC healthcare diagnostics, and environmental monitoring (Singh, 2014, 2016a, b).

2.3.5 Surface plasmon-enhanced fluorescence spectroscopy (SPFS) based detection of agricultural toxins

Haque et al. (2019) have utilized SPR sensors with microchannel incorporated photonic crystal fiber (PCF) to detect low refractive index at near-infrared wavelength. This was achieved by placing plasmonic gold (Au) material and sensing material outside the fiber. A thin layer of TiO_2 was used to attach Au with silica glass. The sensor output was maximum wavelength sensitivity and resolution, which is known to be highest for the PCF-based SPR sensor. This type of PCF-based SPR sensor could be used for a wide range of refractive index detection with higher accuracy for its use in pharmaceutical investigation, leakage monitoring, and biosensing.

2.3.6 Biofunctionalized metal nanoparticles (NPs)

The use of biofunctionalized metal nanoparticles (gold and magnetic) in the SPR has enhanced the detection sensitivity. This has resulted in the improved diagnosis and monitoring of disease biomarkers in different diseases and malignancies. Usually, these assays utilize NPs, which are covalently modified with the analyte molecules such as, antibody (Ab) or a nucleic acid such as DNA/RNA aptamer (Ap) capable of interaction with the related biomarker(s). This results in the increased density near the gold sensing surface and plasmonic coupling with gold film and NPs, which provides a large shift in the refractive index thus enhancing the plasmonic resonance (Fathi et al., 2019). Hassani and Kreysing (2019) have demonstrated the objective-based scanning SPR microscopy to analyze the refractive index of cellular organelles in a noninvasive and label-free manner with a lateral resolution at the limit of diffraction. This could be utilized for investigating the nanometer- to micrometer-scaled specialized sub-cellular organelles reflective of the state of the cell without disturbing the cellular milieu.

2.3.7 AuNPs/GO and AuNPs/GCE hybrid based electrochemical immunosensors

Wang et al. (2017) have developed a sandwich electrochemical aptasensor for the label-free and selective detection of breast cancer cells (MCF-7). This aptasensor utilized a polyadenine (polydA)-aptamer modified gold electrode and a polydA-aptamer functionalized AuNPs/GO hybrid. The decrease in the $[Fe(CN)_6]^{3-/4-}$ differential potential voltammetry (DPV) peak current in the presence of the sandwich system was employed to quantify the breast cancer cells. The high selectivity was ensured by the use of aptamer, which works against the MUC1 protein, overexpressed in MCF-7 cells. A sandwich assay-based DNA biosensor was developed for the detection of *Mycobacterium* sp. This involved sandwiching the target genomic DNA between a specific probe immobilized on indium tin oxide (ITO) electrode modified with a (3-aminopropyl) trimethoxysilane SAM and detector probe/alkaline phosphatase (AP)-dually labeled AuNPs for signal amplification (Zheng et al., 2012). Another sandwich immunosensor developed for HIV p24 protein involved an HRP-labeled detector antibody and electroplating of AuNPs onto a glassy carbon electrode (GCE) surface by chronoamperometry to enhance the conductivity of the electrode and facilitate the immobilization of a large number of capture antibodies while retaining their biological activity (Campuzano et al., 2017d; Singh, 2017b; Zheng et al., 2016).

2.3.8 AuNPs based sensors for the detection of clinical biomarkers

A study led by the Darwish et al. (2016) has developed a sensitive label-free electrochemical immunoassay for the determination of dengue virus biomarker (NS1). This involved an immunosensing scaffold as ITO electrode modified with aryl-diazonium-cation-derived zwitterionic antifouling molecules (4-sulfophenyl, 4-trimethylammoniophenyl, and 1,4-phenylenediamine) capture antibodies and AuNPs. An electrochemical immunosensor was developed for the detection of human enterovirus 71 (EV71), the causative agent of hand, foot, and mouth disease (HFMD) among children. The EV71 virions from the sample were selectively captured on AuNP coated ITO electrode modified with a specific monoclonal antibody and labeled with magnetic nanobeads modified with the detector antibody and HRP (Hou et al., 2017).

A microfluidic-based electrochemical immunoassay was developed for the multiplexed detection of PSA and IL-6 in serum as a marker for prostate cancer. This involved the off-line capture of analytes by HRP-labeled magnetic particles-antibody bio-conjugates followed by the attachment of the capture antibodies to a chip with 8-electrodes modified with glutathione-decorated AuNPs (Chikkaveeraiah et al., 2011).

2.3.9 Colloidal gold-/silver-based electrochemical immunoassay

Colloidal gold or silver are used in an immunoassay as electron-dense probes in electron microscopy since 1970 (Hayatt, 1989). Gonzalez-Garcia et al. (2000) and subsequently Dequaire et al. (2000) have pioneered the use of colloidal gold for the voltammetric monitoring of protein interactions. Costa-Garcia did the same for streptavidin-biotin interactions via absorptive voltammetric measurements. Dequaire et al. (2000) did electrochemical immunoassay via anodic stripping voltammetry (ASV). This protocol was evaluated for a noncompetitive heterogeneous assay of IgG and further utilized for the detection of *Schistosoma japonicum* antibodies (SjAb) in rabbit serum (Yu et al., 2006). Ambrosi et al. (2007) have developed gold nano-labels for optical and electrochemical immunoassay. The sensitivity of AuNP label-based electrochemical immunoassay was enhanced by NP-promoted precipitation of AgNP/AuNP to amplify the transduction of antibody-antigen bio-recognition events. (Chumbimuni-Torres et al., 2006) developed a potentiometric immunosensor based on AuNP labels utilizing sandwich assay where target antigen is captured by primary antibody modified by Au. This was followed by a secondary antibody conjugated with AuNP and Ag and oxidative dissolution of silver with hydrogen peroxide. This has yielded electrolytes for the potentiometric detection of silver ions with a polymer membrane, silver, and an ion-selective electrode. Velev and Kaler (1999) developed a conductive immunosensor having antibody functionalized latex and microelectrode gap. This sensor platform was found to be capable of ultrasensitive detection of IgG down to picomolar level concentration and can be further miniaturized for on-chip protein assays.

2.3.10 AuNP based quartz crystal microbalance (QCM) immunosensing

A novel procedure for dendritic amplification was developed for quartz crystal microbalance (QCM) immunosensing. This protocol used antibody functionalized AuNP as a primary amplifying probe, a dendritic-type immunocomplex of protein A and an antibody-modified Au nanoparticles as the secondary amplifying probe (Chu et al., 2006). Mao et al. (2006) have developed a method where the cyclic accumulation of AuNP is utilized for the detection of IgG using anodic stripping voltammetry (ASV). Zhou et al. (2016) have devised an electrochemical immunoassay based on AuNP labeled immunocomplex enlargement. AuNP label used as electrocatalyst was also utilized to develop an electrochemical immunosensor. Each biorecognition event resulted in a distinct voltammetric peak, the position and size of which corresponds to a particular antigen.

2.3.11 PANIAuNPs based Impedimetric sensor

An impedimetric DNA-based sensor was developed for determining the variable concentrations of *Leishmania infantum* from canine serum samples.

The group had immobilized a specific DNA probe onto a gold electrode already coated with a layer of 3-mercaptopropyltrimethoxysilane on a polyaniline matrix containing AuNPs (PANIAuNPs). Isothermal recombinase polymerase amplification (RPA) with double-labeled amplicons (MBs and AuNPs) along with chronoamperometric detection at screen-printed carbon electrode (SPCEs) was utilized for the detection of *Leishmania* DNA (Campuzano et al., 2017d; Garcia et al., 2016) (Fig. 3).

2.3.12 Multiplexed magneto-immunosensor

Multiplexed sensors with stripping analysis of AgNPs are available. However, their limited stability in saline buffers and tedious synthetic protocols are big deterrents. A multiplexed sandwich-type magneto-immunosensing platform utilizing dual screen-printed carbon electrodes (SPCEs) was developed for the detection of estrogen and progesterone receptors in cell-lysates as cancer

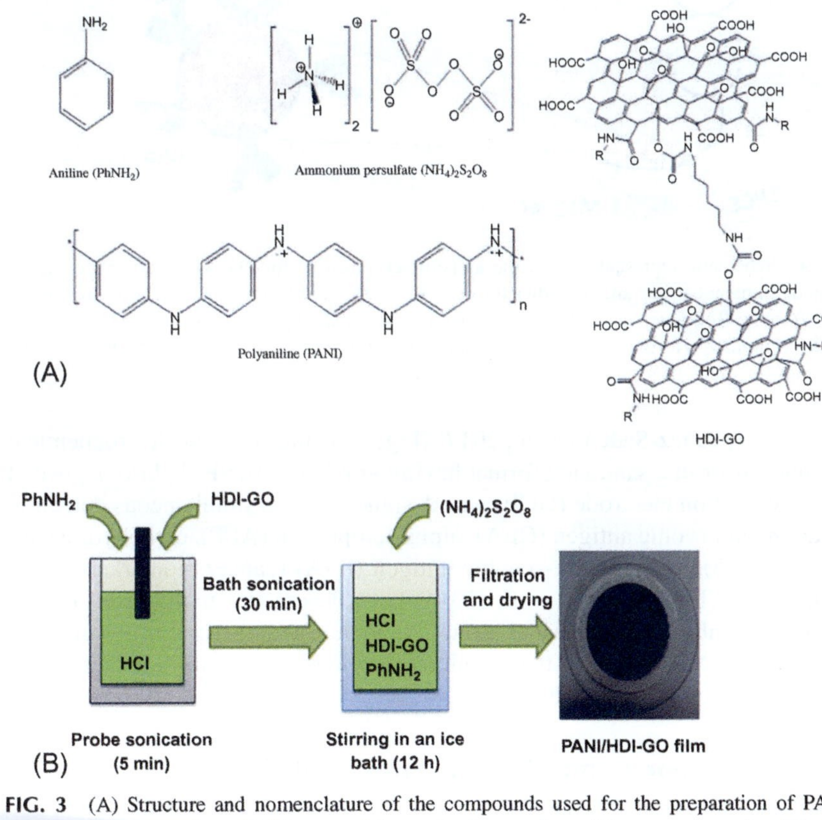

FIG. 3 (A) Structure and nomenclature of the compounds used for the preparation of PANI/HDI-GO nanocomposites. (B) Diagram of the synthesis procedure of the nanocomposites (Sánchez et al., 2019). This is an open-access article distributed under the Creative Commons Attribution License which permits unrestricted use, distribution, and reproduction in any medium provided the original work is properly cited.

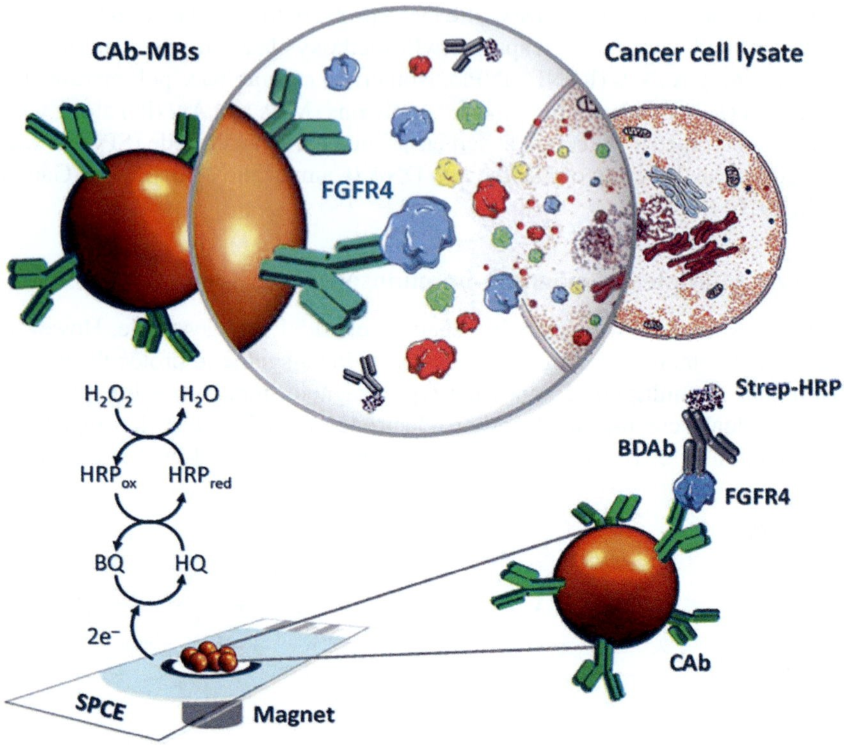

FIG. 4 Schematic representation of the amperometric sensor for FGFR4 determination using magnetic immunocarriers and a sandwich format (Torrente-Rodríguez et al., 2017). Copyright: © 2017 Torrente-Rodríguez et al. This is an open-access article distributed under the terms of the Creative Commons Attribution License, which permits unrestricted use, distribution, and reproduction in any medium, provided the original author and source are credited.

biomarkers (Yáñez-Sedeño et al., 2017) (Fig. 4). A multiplexed electrochemical immunosensor in a sandwich format having graphene-AuNPs hybrid deposited on glassy carbon electrode (GCE) was designed for the simultaneous detection of carcinoembryonic antigen (CEA), alpha-fetoprotein (AFP), carbohydrate antigen (CA125), and prostate-specific antigen (PSA) (cancer biomarkers) (Zhu et al., 2015) (Fig. 5). Yang et al. (2014) have designed a multiplexed immunosensor combining multi-label strategy and multi-spot assay with an array electrode (AuNP/GCE) for the simultaneous detection of six biomarkers for hepatocellular carcinoma (HCC).

2.3.13 Mesoporous metallic structures as labels for electrochemical immunoassays

Mesoporous metallic structures possess interconnected hollow channels that allow mass transport and enhance conductivity. Mesoporous PtNPs (MPtNPs)

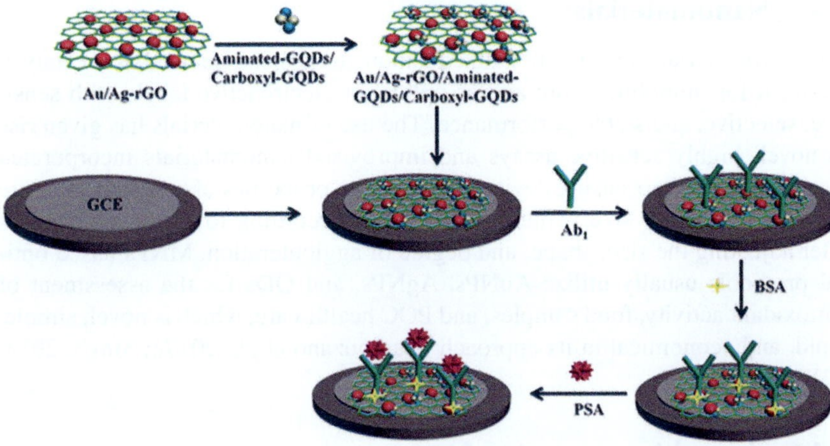

FIG. 5 The illustration of the ECL immunosensor fabrication process (Wu et al., 2016a, b).

acts as nonenzymatic labels for secondary antibodies in sandwich-type immunosensor array. Such platform was utilized for the detection of carbohydrate antigen 125 (CA125), CA153, and carcinoembryonic antigen (CEA) as tumor markers utilizing triple graphene-modified-SPCE (Cui et al., 2014). Graphitized mesoporous/chitosan modified glassy carbon electrode (GCE) was used for the assessment of polyphenols in tea using flow injection analysis (FIA) and an electrochemical detector.

2.3.14 Metal phosphate NP labels

Metal phosphate NP labels are used for the detection of tumor necrosis factor (TNF-α). Metal phosphate labels are easy to prepare and functionalize, requiring only simple diffusion and pH adjustment. Releasing metals from NP labels can be performed in mild conditions (pH 4.6 in acetate buffer) than strong acidic conditions (1 M HNO_3). Metal phosphate NP labels offer a lower detection limits than semiconductor NP labels. These NP labels can be utilized for voltammetric assay of multiple target proteins, yielding a large number of recognizable voltammetric signatures. Each nanoparticle yields a typical multipeak voltammogram, the peak potential and current intensity of which corresponds to a particular biomolecular target (Ding et al., 2013; Li et al., 2015).

2.4. Nanomaterials

The nanomaterials and functionalized nanomaterials can be used as a catalyst; platform for immobilization; and as optical or electroactive labels with sensitive, selective, and stable performance. The use of nanomaterials has given rise to novel, highly sensitive assays and improvised nanomaterials incorporated electrochemical/biosensing devices. The unique properties of nanomaterials are dependent on their size, which can be tuned according to the specific needs after adjusting the size, shape, and degree of agglomeration. MNPs based optical protocols usually utilize AuNPs, AgNPs, and QDs for the assessment of antioxidant activity, food samples, and POC health-care, which is novel, simple, rapid, and economical in its approach (Campuzano et al., 2017c; Singh, 2014, 2016a, b) (Fig. 6).

2.4.1 Au and Ag nanomaterials

Nanomaterials such as AuNP, CNT incorporation offers large surface area, multiple binding sites, conductivity and biocompatibility, sensitivity, selectivity, prolonged stability, and minimum interference from biological matrices hence preferred for the design of affinity-based electrochemical biosensors (Cao et al., 2014; Luna et al., 2015; Peng et al., 2015). Au nanomaterial is used as NPs, nanorods, nanowires and offers good conductivity, strong adsorption,

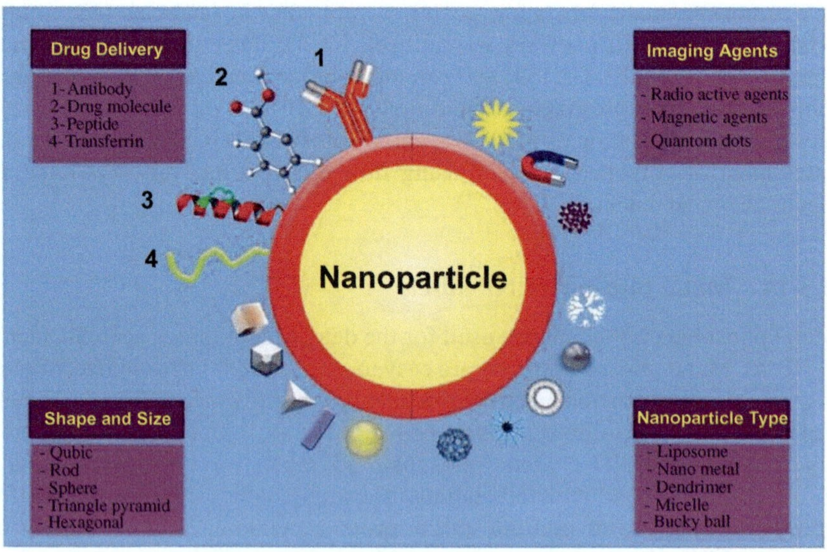

FIG. 6 Schematic representation of nanoparticle types, shapes, and sizes, as well as drug delivery and imaging agents (Barabadi et al., 2019).

biocompatibility and serves as a good nanomaterial for fabricating improvised biosensing platforms. AuNPs exhibits high stability and uniform particle size and is easy to synthesize using a simple protocol. Bifunctional ligands like p-thiophenol, 1,6-hexanedithiol, and polythionine are used to immobilize Au-nanomaterials on the electrode surface through physical or chemical adsorption of capture probes. This is followed by the linking of antibody recognition molecules and finally fixation to the transducer surface. AuNP is used to label secondary antibodies to functionalize multiwalled carbon nanotubes (MWCNTs) for antibody labeling, nano-porous gold (NPG), and nanoclusters before the immobilization of antibody for the highly sensitive detection of the analyte (Cao et al., 2014; Luna et al., 2015; Peng et al., 2015).

2.4.2 Au and ag-based electrochemical immunosensor

Wang et al. (2015) have designed AgNP-labeled CEA immunosensor and further Au-Ag core-shell NPs for the detection of platelet-derived growth factors. (Chu et al., 2006) have reported an AgNPs colloidal Au-labeled immunosensor. Bai et al. (2013) have developed a sensitive and selective aptamer/thrombin/aptamer-AuNPs sandwich enhanced SPR sensor for the real-time detection of sub-nanomolar thrombin. The study involved immobilizing the thiol-modified thrombin aptamer (TBA29) on AuNPs via Au-S bonding. The other biotinylated thrombin aptamer (TBA15) was attached to the streptavidin pretreated SPR gold film through biotin-streptavidin recognition leading to the formation of a double aptamer sandwich structure, thus enhancing the SPR signal, which is directly proportional to the thrombin concentration. This protocol took advantage of sandwich binding of two affinity aptamers for increased specificity, AuNPs for signal enhancement, and SPR signal readout for real-time detection. The SPR signal was found to be in a good linear relationship with thrombin concentration. The presence of AuNPs enhanced the LOD 5-fold and also the selectivity as it was not affected by the presence of other proteins like BSA and lysozyme. This SPR sensing platform was utilized for analyzing the thrombin in diluted human serum samples (for references Pan et al., 2017; Singh, 2014, 2016a, b).

2.4.3 Nanomaterial-based enzyme electrodes for the estimation of polyphenols

Enzyme electrodes have great utility in biosensing applications owing to the analytical potential of coupling selectivity and catalytic action. Nanomaterials provide unique and creative configurations for enzyme electrodes. Enzyme electrodes for polyphenols are usually made up of polyphenol oxidase (PPO, tyrosinase, catechol oxidase, and laccase) having copper as a prosthetic group and molecular oxygen as co-substrate. Ziyatdinova et al. (2008) have designed MWCNTs-DNA based biosensors for estimating polyphenols in rutin and tea extracts.

A laccase-MWCNT-chitosan-based biosensor was developed for the total phenol estimation in vitro in *Salvia officinalis* and *Mentha piperita*. SPE-SWCNTs modified electrodes and polyazetidine were used to immobilize laccases from *Trametes versicolor* and *Trametes hirsute* while another group has designed gold electro-C-MWCNT-polyaniline/AgNPs for the immobilization of laccase. Another study has substituted AgNPs with NiNP and got the same results in terms of sensitivity, and stability for the estimation of polyphenols. Graphene oxide (GRO) integrated with MWCNTs were used as tyrosinase and laccase enzyme electrodes for phenol estimation in fruit juices. GRO along with MWCNTs in a 1:1 ratio improves the performance of GCE. One study has used multilayered enzyme electrodes having AuNPs, fullerene, and *Trametes versicolor* laccase as components on Au-electrode surface. Although the electrode gave a slightly higher estimation of polyphenols when compared with FC, it retained the activity even after 120 reactions. A tyrosinase carbon black paste electrode (CBPE) was developed for total phenol determination in olive oil. Results show the detection of 20 different PCs from olive oil (Bergwerff and van Knapen, 2006; Blasco et al., 2007; Manikandan et al., 2018).

2.4.4 CID-LSPR using gold nanorods (AuNR) for bio-detection

Moon and Ha (2019) have devised a chemical interface damping (CID) based localized surface plasmon resonance (LSPR) biosensor to detect biological molecules having thiol or sulfur in a real-time manner. The detection was free from any interference from the dielectric constant in single AuNRs. The study has utilized scanning electron microscopy (SEM) correlated dark-field (DF) scattering studies of gold nanorods (AuNRs) at the single-particle level to understand CID as well as alternative functional groups other than thiol groups (-SH, sulfhydryl groups) that can induce CID. Results show that biotin along with sulfur can cause strong plasmon damping in single AuNRs (Singh, 2016a, b).

2.4.5 Other metal nanomaterials (Cu, Pd, Pt)

In addition to Au and Ag nanomaterials, Cu, Pd, Pt, and others are also used as nanomaterials for the fabrication of novel electrochemical immunosensor with high detection sensitivity (Jiao et al., 2016). Zhang et al. (2014) have used Cu-doped titanium dioxide nanoparticle (TiO_2-NP) with high electrocatalytic activity as a label to construct sandwich-type immunosensor. Pt nanoparticles (PtNPs) come in two different morphological forms, i.e., hollow and solid Pt-nanospheres displaying variable electrochemical properties to be used as labels. Cui et al. (2014) have used mesoporous PtNPs (M-Pt-NPs) as nonenzymatic labels for designing disposable immunosensor arrays. In a study, signal tags with distinct electrochemical behavior were prepared by complexing Cd^{2+} and Cu^{2+} with amino-capped platinum porous nanoparticles (PtPNPs) followed by conjugation with anti-CEA and anti-AFP antibodies, respectively, which are used as distinguishable signal tags for capturing antigens. A GCE modified with an ionic liquid was used as the electrode substrate for the immobilization of capture antibodies. The voltammetric signals monitored by DPV corresponds to the concentration of antigen (Yáñez-Sedeño et al., 2017).

2.4.6 Carbon-based nanomaterials

Carbon-based nanomaterials such as carbon nanotubes (CNTs), Graphene, Graphene oxide (GO), carbon nanofibres, carbon nanospheres, bucky-paper, etc. offers good conductivity, biocompatibility, robust mechanical strength, and large surface area for increased immobilization of biomolecules and improved active sites for electrochemical reactions (Pan et al., 2017)' (Figs. 7 and 8). The ability of carbon to hybridize sp., sp2, and sp3 configurations with narrow gaps between their 2 s and 2p orbitals impart versatility in the design of carbon-based nanomaterials for the sensitive detection of biological analytes (Singh, 2016a, b; Wanekaya, 2011). Hydrothermal carbonization of carbohydrates gives rise to

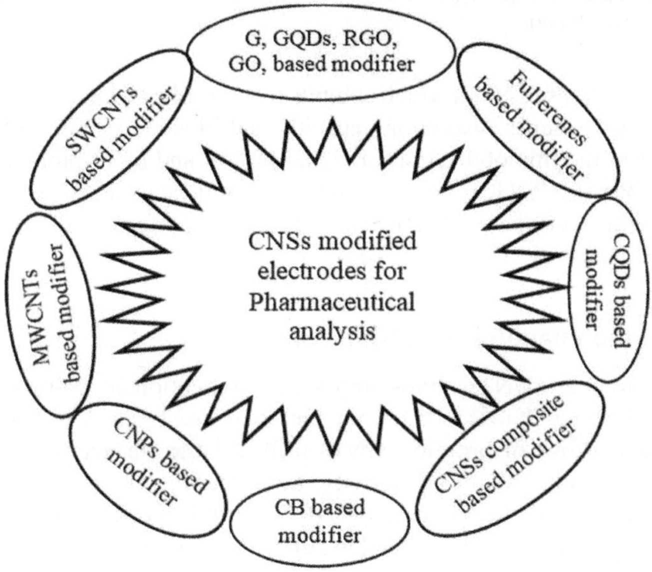

FIG. 7 Various carbon nanostructured based modified electrodes (*CNSs*, Carbon nanostructures; *SWCNT*, Single-walled carbon nanotubes; *G*, Graphene; *GQDs*, Graphene quantum dots; *GO*, Graphene oxide; *RGO*, Reduced graphene oxide; *CQDs*, Carbon quantum dots; *CB*, Carbon black; *CNPs*, Carbon nanoparticles; *MWCNTS*, Multi-walled carbon nanotubes) (Ghalkhani and Ghorbani-Bidkorbeh, 2019). Copyright notice. This is an Open Access article distributed under the terms of the Creative Commons Attribution License, (http://creativecommons.org/licenses/by/3.0/) which permits unrestricted use, distribution, and reproduction in any medium provided the original work is properly cited.

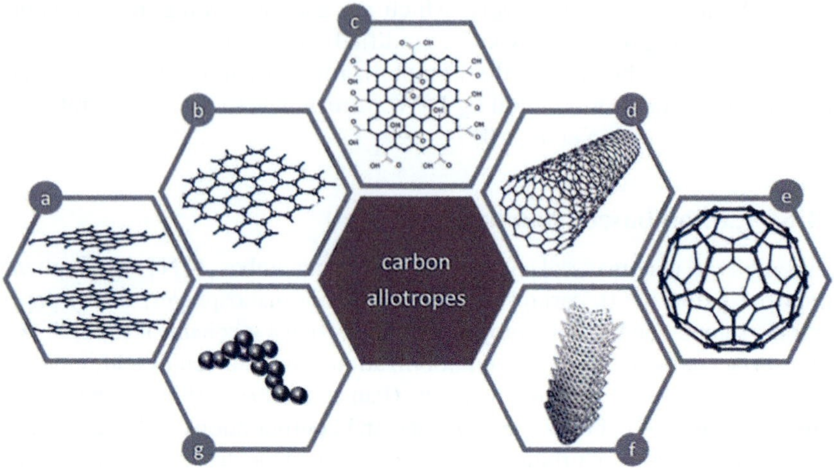

FIG. 8 Carbon allotropic forms: (A) Graphite, (B) graphene, (C) graphene oxide, (D) carbon nanotube, (E) fullerene, (F) carbon nanofibers, (G) carbon dot (Cardoso et al., 2019).

carbon nanospheres (CNSs), which exhibits features like abundant reactive oxygen functional groups, adsorption capacity, and biocompatibility. CNNs find application in the immobilization of signal probes and other biomolecules. A CNSs-AgNPs/AuNPs/GCE-AuNPs/rGO nanocomposite was utilized for fabricating a multiplexed electrochemical immunosensor for the simultaneous detection of CEA and AFP (Li et al., 2016a).

2.4.7 Carbon nanotubes (CNTs)

Carbon nanotubes (CNTs) finds important application in fabricating electrochemical sensors/biosensors. It enables the investigation of the electron transfer capabilities of proteins. CNTs offer a large surface area to weight ratio, and good electrical conductivity, thus finding important applications in the fabrication of immunosensors. CNTs find application as electrode modifying substrate due to intermolecular electron transfer, and as markers in sandwich-type immunosensor (Yu et al., 2006). Wang et al. (2004) have employed CNTs as a career for covalently bonding a large number of enzyme molecules. Liu et al. (2012) have used SWCNTs for immobilizing antibodies on GCE. Jeong et al. (2013) have used MWCNTs as dendritic nanostructure matrix for immobilizing antibodies for the CEA immunosensor. CNTs offer excellent 1D conductivity, and high-aspect-ratio resulting in an increased surface area of the electrode. CNTs can be combined with Pt-NPs to form nanocomposites, which leads to improved detection sensitivity due to increased surface area of CNTs and catalytic properties of Pt (Zhu et al., 2005). CNTs can also combine with AuNPs resulting in increased rates of electron transfer between the enzymatically generated substrate and CNT-AuNP composite (Holzinger et al., 2017).

2.4.7.1 CNTs: Clinical biomarker detection

Methylglyoxal, a biomarker for type 2 diabetes mellitus indicating intima-media thickening, vascular stiffening, and the elevation of systolic blood pressure was detected in the plasma of diabetic patients using SWCNTs (Adhikari et al., 2015). A study led by the Wang group has devised ultrasensitive electrochemical immunoassay where CNTs were utilized as careers to load enzyme tracers (9600 enzymes/CNT of 1 μm length). An enzyme-CNT label resulted in a 100-fold enhancement in the electrochemical signal and detection sensitivity in the attamolar range. A CNT forest amperometric immunosensor platform was developed for the detection of cancer biomarkers in serum and tissue lysates utilizing multi-labeled secondary antibody-CNT bioconjugates (Adhikari et al., 2015). A study led by Gayathri et al. (2016) have utilized a glassy carbon electrode (GCE) scaffold modified with thionine dye immobilized chitosan/functionalized-MWCNT composite (GCE/f-MWCNT-Chit@Th) for the detection of uropathogenic *Escherichia coli* (UPEC) (Campuzano et al., 2017d). A multiplexed immunosensing platform was developed utilizing dual SPCEs modified with double-walled CNTs for the simultaneous detection of TNF-α and IL-1β cytokine biomarkers (Sánchez-Tirado et al., 2017). An immunosensing platform was developed for the detection of CA15-3 that does not require reagent and mediator. This platform utilized GCE modified with CNTs and nanospheres with bilayered Pt nanoclusters for glucose oxidase (GOD) and adsorbed silica. Results show that peak current corresponding to cofactor oxidation decreases with the increasing concentration of CA15-3.

2.4.7.2 CNTs: Detection of polyphenol

Carbon nanotubes (CNTs) are equipped with unique structural, mechanical, and electronic properties permitting their role in different sensing protocols. Glassy carbon electrode (GCE) surfaces modified with single-walled carbon nanotubes (SWCNTs) can be used for the detection of polyphenols or the estimation of total polyphenols in wine using multi-walled carbon nanotubes (MWCNTs) and measurement in flow injection analysis (FIA). Similarly, GCE modified with SWCNTs and covalently functionalized with polytyrosine (GCE-SWCNT/Polytyr) nanocomposites were used for the determination of total polyphenolic content in tea extracts. An integrated SPE-MWCNTs with a microfluidic platform (glass "double T" microfluidic chip with a 74 mm long separation channel) was utilized for the food polyphenols analysis. The use of MWCNTs improves the selectivity by increasing the number of theoretical plates and resolution as fast electron transfer sharpens the peak and reduces their broadening (Felix and Angnes, 2018).

2.5. Graphene-based nanomaterials

Graphene is a carbon material, where a 2D-sheet of carbon is linked by sp^2 bonds. This configuration endows graphene with extraordinary properties such as high conductivity, electron mobility at room temperature, low energy dynamics of electrons with atomic thickness, robust mechanical behavior, and flexibility making it invaluable for electrochemical and analytical applications (Liu et al., 2007). Graphene displays unique physiochemical properties like monolithic structure, high conductivity, large specific surface area, no toxicity, and good electron mobility, which facilitates its use in electrochemical biosensing. Graphene-modified electrodes have been successfully used for the detection of H_2O_2, NADH, dopamine, ascorbic acid, uric acid, and acetaminophen. Graphene is utilized for the modification of GCE and antibody; for the development of sandwich immunosensor by getting reduced on SPE surface; as immune-probe (graphene/AuNP) for the detection of CEA; and serves as a carrier of thionin and HRP in a sandwich immunosensor (Pan et al., 2017). Cetó et al. (2012) have devised a bio-electronic tongue system made up of an array of epoxy-graphite modified electrodes for simultaneous detection of different polyphenols utilizing different redox enzymes and CuNPs trapped in epoxy resin.

A wide array of graphene-based nanomaterials such as gold nanoparticles, mono- and bimetallic nanoparticles and metal nanocomposites, etc. are used for the design and synthesis of high-performance electrochemical sensing devices with high sensitivity, selectivity for medical-diagnostics, environment biomonitoring, and food safety monitoring in a real-time manner (Cardoso et al., 2019).

2.5.1 Graphene oxide (rGO)

Different variants of graphene like graphene-oxide (rGO) are prepared by introducing oxygen-containing functional groups or reduced graphene, which is prepared by incomplete removal of functional groups by chemical or heat treatment. rGO and reduced graphene have been used in immunosensors for the simultaneous determination of CEA and AFP. AuNP incorporated CNT-graphene composite and functionalized mesoporous materials (MCM-41) were utilized for enhanced detection sensitivity (Pan et al., 2017). Interaction of monolayer graphene with surface plasmons on the Au-surface enhances the optical sensitivity of SPR sensors as observed in the case of the anti-cholera toxin SPR sensing platform (Singh et al., 2015). Electrochemical sensors and biosensors based on 2D-graphene materials like boron nitride (BN), graphite-carbon nitride (g-C_3N_4), transition metal dichalcogenides (TMDs), transition metal oxides, and graphene are also available. These are better suited for the detection of important biomarkers like glucose, hydrogen peroxide, and cancer biomarkers (Bollella et al., 2017).

2.5.2 Ultrathin graphitic carbon nitride (g-C3N4) nanosheets

g-C3N4 nanosheets with their 2D-graphene-like structure and unique physicochemical properties have attracted great research interest (Fig. 9). Recently,

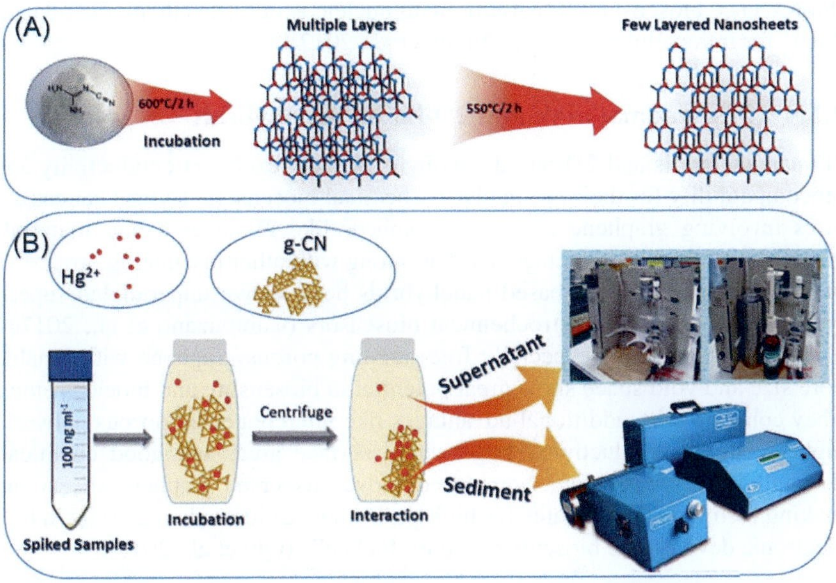

FIG. 9 Schematic representation of the synthesis of nanosheets using the high-temperature short-time (HTST) polycondensation process (Chouhan et al., 2019). This is an open-access article distributed under the Creative Commons Attribution License which permits unrestricted use, distribution, and reproduction in any medium provided the original work is properly cited.

proton-functionalized ultrathin g-C3N4 nanosheets were prepared by sonication-exfoliation of bulk g-C3N4 under acidic conditions. This method reduces the exfoliation time from more than 10h to 2h and endows the nanosheets with positive charges. Besides retaining the properties of g-C3N4, the obtained nanosheets with the thickness of 2–4nm (i.e., 6–12 atomic monolayers) also exhibit a large specific surface area of $305\,m^2 g-^1$, enhanced fluorescence intensity, and excellent water dispersion stability due to their surface protonation and ultrathin morphology. The well-dispersed protonated g-C3N4 nanosheets facilitate interaction with negatively charged heparin resulting in the quenching of g-C3N4 fluorescence. A highly sensitive and highly selective heparin sensing platform based on protonated g-C3N4 nanosheets was developed. This metal-free and fluorophore label-free system gave the lowest heparin detection limit of $18\,ng\,mL^{-1}$ (Ma et al., 2014).

2.5.3 2D-graphitic carbon nitride nanosheets (CNNSs)

Carbon nitride nanosheets (CNNSs) with planar graphene-like structures show superior stability, high fluorescence quantum yield, low toxicity, excellent biocompatibility, unique electroluminescent, and photoelectrochemical properties, which make them suitable candidates for biosensing. CNNSs have great utility in photoelectrochemical biosensing, electrochemiluminescence biosensing, and fluorescence biosensing. Integration with nucleic acid makes them a good candidate for intracellular imaging (Xiong et al., 2017).

2.5.4 2D-nanomaterials and 2D-based nanohybrids

2D-nanomaterials and 2D-based nanohybrids show excellent conductivity and biocompatibility for their use in electrochemical biosensors. Hybrid nanostructures involving graphene and other graphene-like 2D-layered nanomaterials (GLNMs) are highly advantageous. This along with other fascinating properties of 2D-nanomaterials, 2D-based nanohybrids have shown unparalleled superiorities in the field of electrochemical biosensors (Campuzano et al., 2017c). 3D-graphene materials especially free-standing porous graphene with tunable pore size and void space show great potential in biosensing and bioelectronics. They come up with additional advantages like interconnected porous network, high electronic conductivity, large active surface area, and good chemical/thermal stability, providing them an edge over-dispersed graphene sheets and making them good candidates for high-performance electrode materials in bioelectronic devices like biosensors and biofuel cells (Qiu et al., 2017).

2.5.5 Graphene (G)-based composite materials

Graphene nanocomposites have a major role in sensing platforms due to their atom-thick 2D-conjugated structures, high conductivity, large specific surface area, and controlled modification. With the enormous advantages of film structure, G-based composite films (GCFs) can be prepared by combining graphene with different functional nanomaterials (noble metals, metal compounds, carbon materials, polymer materials, etc.). GCFs show unique optical, mechanical, electrical, chemical, and catalytic properties resulting in sensors with high sensitivity, selectivity, and stability (Yu et al., 2017). A multiplexed immunosensor using dual SPCE modified with reduced graphene oxide (rGO) was used for the detection of PYY and GHRL in biological fluids for diabetes (Verma et al., 2011). Zhou et al. (2016) have constructed a simple, dual-, single-labeled electrochemical immunosensor using rGO as the substrate deposited onto a GCE to immobilize the antibodies. The signal probes used in this configuration were CdSeTe@CdS quantum dots and Ag nanoclusters (AgNCs) incorporated into mesoporous silica nanospheres for the detection of B-cell lymphoma-2 (Bcl-2) and Bcl-2 associated X protein (Bax) as a biomarker for monitoring the apoptosis of tumors.

2.5.6 Graphene-based chronoamperometric genosensor for bio-detection

Another sandwich-type electrochemical chronoamperometric genosensor was employed for the detection of BRAC1, a cancer biomarker utilizing rGO/GCE

(Campuzano et al., 2017c). Esteban-Fernández de Ávila et al. (2015) has designed an electrochemical DNA biosensor utilizing SPCEs modified with a reduced graphene oxide-carboxymethylcellulose hybrid nanomaterial for the detection of a target DNA sequence of the *p53* tumor suppressor gene. Human papillomavirus (HPV) was detected using an electrochemical genosensor in hybridization format by immobilizing a specific capture probe on a GCE modified with graphene/Au nanorods/polythionine. Similarly, a genosensor was designed for the detection of miRNA as a biomarker for cancer and cardiovascular disease. The protocol involved direct or competitive hybridization assay using RNA-RNA, DNA-DNA duplexes as specific recognition sequences. Shuai et al. (2016) have developed an ultrasensitive electrochemical biosensor for miRNA-21 (a biomarker involved in cancer initiation). This was based on tungsten oxide-graphene composites (WO_3-Gr) coupled with catalyzed hairpin assembly target recycling and enzyme signal amplification.

2.5.7 Gold-graphene nano-labels for the detection of cancer biomarkers

An electrochemical immunosensing platform was developed for the detection of carcinoembryonic antigen (CEA) utilizing a magnetic nanocomposite with redox activity and gold-graphene nano-labels. Gold-graphene nano-labels were used for facilitating AuNP immobilization and multiple biomolecules, permitting antigen-antibody interaction (Campuzano et al., 2017a, b, c, d).

2.5.8 Graphene-based electrochemical CEA immunosensor

A label-free electrochemical CEA immunosensor was developed that utilized graphene for good electrical conductivity and bioactive surface area. It also contained AuNPs for biocompatibility and chitosan for good adhesion, nontoxicity, and abundant reactive amino groups to facilitate the synthesis of a cryogel for better electron transfer. A nanostructured nonenzymatic electrochemical immunosensing platform was designed for CEA. This platform utilized GCE modified with a nanocomposite of stannic oxide/reduced graphene oxide (SnO_2/rGO) and AuNPs for immobilizing capture antibodies. Nanocomposites facilitated a large surface area for electron transfer and antibody immobilization. The captured CEA was then used with PdNPs-vanadium pentoxide (V_2O_5)/MWCNTs linked with secondary antibodies (Ab_2). The PdNPs-V_2O_5/MWCNTs were used for the catalytic reduction of H_2O_2 (Campuzano et al., 2017b). A sandwich-type electrochemical CEA immunosensor was proposed utilizing an enzyme-free signal amplification strategy aided by graphene oxide/chitosan-ferrocene (GO/CS-Fc) as a matrix for immobilizing capture antibodies and Fe_3O_4/AuNPs functionalized detector antibodies. CEA was determined by the catalytic reduction of p-nitrophenol to electroactive p-aminophenol using gold-nanocatalyst labels (Peng et al., 2016).

2.5.9 Magnetic graphene-based electrochemical CEA immunosensor

A flow-through multiplexed immunoassay in sandwich format was designed for the simultaneous electrochemical detection of CEA and AFP utilizing magnetic graphene-nanosheets with immobilized capture antibodies as immunosensing probes, and multifunctional nanogold hollow microspheres (GHS) with encapsulated HRP-THI and HRP-Fc as distinguishable signal tags (Yáñez-Sedeño et al., 2017).

2.5.10 Graphene-based Sandwich immunoassay for the detection of cancer biomarker

Sun et al. (2015) have constructed a microfluidic paper-based sandwich immunoassay using (μ-PAD) in which rGO was used to modify the cellulose fibers to determine HCG, PSA, and CEA. ZnO-nanorods with a large surface-to-volume ratio and abundant sites for capturing biomolecules were employed to immobilize the specific antibodies.

2.6. C_{60} fullerenes and carbon dots

Carbon dots or carbon quantum dots, sometimes also called graphene quantum dots/C_{60} is a nondimensional, carbon nano-object utilized as redox mediator and nano-scaffolds in enzymatic sensors. It can combine with AuNPs and PtNPs for improved electrochemical, gravimetric, immune, and electroluminescent aptasensing applications (Wanekaya, 2011). Carbon QDs and semiconductor QDs have similar functional efficiency in FRET-based biosensing and as fluorescence labels. Antibody-antigen modified AuNP and carbon QDs exhibits efficient FRET between them (Bu et al., 2014; Wanekaya, 2011). Hemmateenejad et al. (2012) have used L-cysteine-capped CdTe QDs for the quantification of polyphenols. A PAMAM dendrimer-metal sulfide QDs utilized as the signal label in multiplexed sandwich-type electrochemical immunosensor simultaneously detected CA125, CA15-3, and CA19-9 cancer biomarkers employing stripping voltammetry (SV).

2.7. Carbon black

Carbon black (CB) is a carbon material with a spherical particle-based primary structure and aggregates as a secondary structure. It is used as filler for rubber, plastics, and composites. Carbon nanoparticles (CNPs) are synthesized using CB dispersion followed by sonication in appropriate solvents. NanoCB shows excellent conductivity and electrochemical behavior. Electrochemical biosensors employing CB are used for pesticide detection. CB modified with SPE are used for polyphenol detection in foods, while Printex L6 nano-carbon black along with AgNPs are used for the detection of total phenol in wine (Felix and Angnes, 2018).

2.8. Carbon bucky-paper

Carbon buckypaper is a thin film (5–25 μm) that consists of laterally arranged networks of nanotubes used for synthesizing nanocomposites. It has been utilized for the construction of mediator-free glucose sensors with high sensitivity, stability, selectivity, and reproducibility (Ahmadalinezhad et al., 2011) (Fig. 10).

2.9. Other carbon materials

Carbon nanofibers exhibit large functional surface area and numerous surface-active groups. **Carbon nanospheres** display rich functional groups exhibiting good biocompatibility, dispersibility, and activity. These are relatively less toxic than CNTs. Soluble carbon fibers and carbon nanospheres can be utilized for the construction of highly sensitive immunosensor (Pan et al., 2017). **Carbon nanodots** function as a good alternative for QDs for their role in biosensing and imaging. **Carbon paste electrodes (CPE)** can conjugate with other NMs for electrochemical transduction resulting in improved biosensing. A carbon paste electrode (CPE) modified with core-shell hybrid nanomaterials constituted by surface-active maghemite nanoparticles (core) and tannic acid (shell) was utilized for the determination of polyphenols in blueberry extracts. **Glassy carbon, doped diamond**, and **graphite** are standard materials for the synthesis of electrodes for electrochemical biosensing, and other bioanalytical applications (Holzinger et al., 2017).

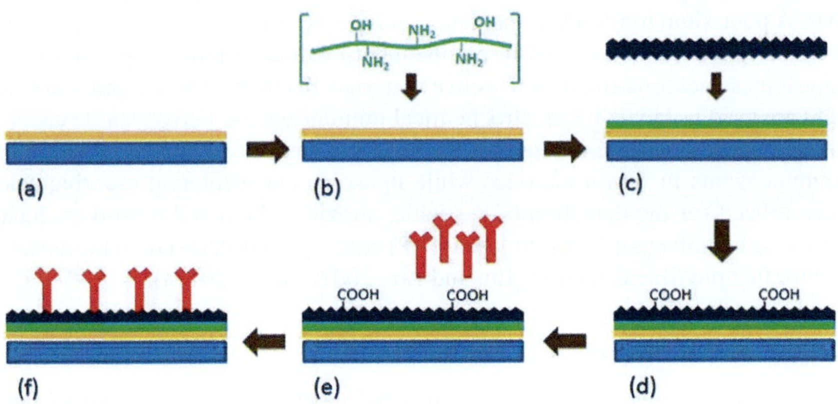

FIG. 10 Schematic diagram of the preparation of the buckypaper-based electrochemical biosensor. (A) a titanium plate sputtered with a thin layer of gold; (B) gold interaction with chitosan; (C) attachment of buckypaper to the gold surface via chitosan; (D) activation of the buckypaper; (E) immobilization of enzymes; (F) the fabricated biosensor (Adhikari et al., 2015).

2.10. Apoferritin nano-vehicles

Apoferritin is the new entrants among the NP labels for the ultrasensitive electrochemical immunoassay of proteins (Liu and Lin, 2007). These are spherical protein shells of around 12.5 nm diameter, composed of 24 subunits surrounding an aqueous cavity (8 nm diameter) that can accommodate about 4500 iron atoms. The protein cage of apoferritin dissociates into 24 subunits at low pH (2.0) and reassociates at high pH (8.5). Apoferritin has 6 hydrophobic and 8 hydrophilic channels. Apoferritin labels are of two types, i.e., redox probe loaded apoferritin and metallic phosphate loaded apoferritin. The concept was first demonstrated when hexacyanoferrate was loaded in the cavity of apoferritin for the electrochemical immunoassay of an IgG analyte. Apoferritin template synthesis of the encoded NP provides a quick and simple approach to synthesize electrochemical NP labels for veritable applications (Liu and Lin, 2007; Singh, 2017a).

2.11. Liposome

Liposomes are lipid bilayer structure that comes in the size of a nanometer (nm) to micrometer (μm). The hydrophobic chains of the lipids form the bilayer while the polar headgroups of the lipids orient towards the extravesicular solution and inner cavity. Liposomes serve as a preferred bioassay label owing to the high surface area, large internal volume, and ability to get linked with a variety of biorecognition elements via its lipid layer. A label may incorporate enzymes, fluorescent dyes, electrochemical, and chemiluminescent markers, DNA, RNA, ions, and radioactive isotopes. These labels could be either ion-encapsulated liposome labels or ion-selective electrodes for serum protein analysis. A potassium ion-loaded liposomes and a potassium ion-selective electrode were designed for monitoring the complement-mediated immune-lysis reaction. Liposomes encapsulating electroactive materials like ferrocyanide and ascorbic acid are used as labels for electrochemical immunosensor. Ferrocyanide encapsulated liposomes were utilized for the sensitive and selective determination of immunoagents in serum matrices while liposome encapsulating ascorbic acid was utilized for the detection of pesticide, atrazine. Haga and coworkers have developed a horseradish peroxidase (HRP) encapsulated liposome immunosensor for theophylline detection (Liu and Lin, 2007; Singh, 2017a).

2.12. Semiconductor nanomaterials

Semiconductor nanomaterials like silica NPs and QDs exhibit large surface area, high surface reaction activity, strong adsorption, career for immobilization of proteins and enzymes, easy functionalization due to the large number of unsaturated groups and hydroxyl groups, etc. facilitating their use in biosensor design for enhanced sensitivity (Liu et al., 2011). Hong et al. (2009) have designed a reagent-free immunosensor using ferrocene-coated silica NPs.

Wu et al. (2012) have designed immunosensor utilizing NPG and hollow mesoporous silica nanospheres (HSMs) for enhanced sensitivity owing to the large surface area of amino-functionalized HSMs providing an increased surface for loading antibodies, mediators, and enzymes. Yang et al. (2010) used silica balls/spheres for immobilizing biomolecules for highly sensitive detection of cancer biomarkers.

2.12.1 SiO_2 nanomaterials

Tang et al. (2010) have used thionine-HRP complex into a nano-silica solution for adding to HRP-labeled secondary antibody to construct an electrochemical immunosensor. Similarly, in another approach, uniform dispersion of silica nanospheres was used along with silane functionalized surface for the labeling of signal antibody and loading high levels of HRP enzyme (Wu et al., 2009).

2.12.2 Silica nanoparticles

Semiconductor NPs such as quantum dot (QD), CdS, PbS, and ZnS play important role in biosensing owing to their unique optical properties like size-tunable fluorescence and sensitivity. Silica NPs can be utilized for bio-imaging, optical bioassays, drug delivery, diagnosis, and therapeutic applications. Different semiconductor materials like ZnS, CdS, PbS, CuS, etc. are used to label different proteins such as α2-microglobulin, IgG, bovine serum albumin (BSA), C-reactive protein (CRP) to identify target analytes employing stripping voltammetry (SV). An electroactive silica NP was developed where poly (guanine) (poly [G]) NP was used to functionalize silica NPs. These poly (G) functionalized NPs serve as a label in an electrochemical immunoassay (Wang et al., 2006). Wang et al. (2006) have utilized electroactive silica NP for the detection of TNF-α with a detection limit in attomolar range.

A concept was developed for the electrochemical DNA hybridization assay, which was further improvised to electrochemical immunoassay for the simultaneous measurement of multiple protein targets, i.e., multi-analyte electrochemical immunoassay. The study has utilized a carbamate linker for linking hydroxyl-terminated semiconductor NPs with secondary antibodies (Wang et al., 2006). A similar protocol utilizing QD (ZnS@CdS) labels was developed for the electrochemical immunoassay of interleukin-α. Biorecoginition event was followed by concentration determination by electrochemical stripping analysis of cadmium captured labels after an acid dissolution step. Similarly, a QD/aptamer-dependent ultrasensitive electrochemical biosensor was developed to detect multiple protein targets. This study has utilized a single-step displacement assay involving co-immobilization of several thiolated aptamers along with the binding of corresponding QD-tagged proteins on gold surface. Displacement was monitored through the electrochemical detection of the leftover nanocrystals. This sensing platform has the advantage of being a portable, low-cost, and low-power consuming device. Additionally, as a single-step displacement assay, it scores over the two-step sandwich assay used in QD-based electrochemical immunoassay (Algar et al., 2017). A QD-label was utilized in a potentiometric immunoassay for the bioanalysis of protein in a microtiter plate format. Kuo et al. (2019) fabricated a dielectric-grating-coupled SPR sensor using a low-cost nano-imprint process. This sensor exhibited high phase detection sensitivity when the light is incident on the backside of the sensor and does not pass through the analyte on the front-side of the sensor. This study has utilized a low refractive index material like mesoporous silica to implement a reverse symmetric waveguide structure that can enhance electric-field strength on the sensor surface and improve detection sensitivity. Authors have used the sol-gel method to fill the groove of the grating structure with a high refractive index material (titanium dioxide, TiO_2). Surface smoothness is improved through a flat silicon impression mold. The experimental results show high phase detection sensitivity achieved using an electro-optic heterodyne interferometer. However, the sensor does exhibit defects and nonsmooth surface relief (Kuo et al., 2019).

2.12.3 Quantum dots

QDs are semiconductor nanocrystals that come up with size-dependent fluorescent properties. QDs score over conventional organic fluorescence probes having the advantage of broadband excitation, narrow bandwidth, and high-intensity emission. Recently, the field of analytical chemistry has witnessed a fillip in the use of quantum dots (QDs) as nanomaterials. A variety of QDs combines the characteristic properties of nanomaterials and give sharp and sensitive dissolution peaks through electrochemical dissolution voltammetry (Algar et al., 2017). QDs can be used as electrochemical and fluorescent labels. QDs are semiconducting nanocrystals required for fluorescence-based transduction in DNA or aptamer-based sensing. QDs availability in different wavelengths makes them an interesting candidate for multiplexed analysis (Malic et al., 2011; Moro et al., 2017). QDs can be combined with AuNPs to increase the excitation rate of QDs and fluorescence with magnetic NPs for improved biodetection. QDs can also interact with propagating surface plasmons leading to enhanced signal output in SPR sensing as observed in the case of ss-DNA and prostate-specific antigens. QDs in conjunction with CNTs form domains with variable conductivity, behaving like a microelectrode array (Malic et al., 2011; Moro et al., 2017). QDs have important applications as fluorophores in biological applications due to the size-tuned fluorescent emission as a function of the core size, shape, and material. Graphene-based QDs (GRQDs) were utilized for the polyphenol evaluation in olive oil (Malic et al., 2011; Moro et al., 2017).

Liu et al. (2019) have developed immunochromatographic strip (ICS) assay based on quantum dots (QDs) for qualitative, quantitative, rapid, and sensitive detection of acetamiprid in agricultural products. The competitive immunoassay utilized a QDs-mAb conjugated probe. The protocol was rapid and found to be free from cross-reactivities with another neonicotinoid. The accuracy and precision were very high. With other analytes also, QDs-ICS showed comparable reliability and high correlation with ultra-performance liquid chromatography-tandem mass spectrometry. A similar protocol could be applied for POC testing of other analytes.

2.13. Other nanomaterials

Magnetic microspheres are magnetic-nanomaterial displaying magnetic response and nano-synergistic effects showing characteristic features of magnetic microspheres. **Magnetic gold nanospheres** were used along with thionin-labeled HRP and Prussian blue labeled HRP oxidase to prepare CEA immunosensor. It was also used for the detection of prostate antigens via carboxylated magnetic nanospheres or NPs-functionalized phosphopeptide nanospheres for simultaneous detection of AFP, cardiac troponin I, and human myocardial fatty acid-binding protein in a sandwich immunosensor format. Many of these nanomaterials show synergetic effects with other nano-objects, for instance, nonradiative energy transfer using upconverting nanoparticles (UCNPs) and quantum dots (QDs) (Li et al., 2016b).

2.14. Other labels

In addition to their use as nanomaterial labels, micrometer size materials such as polystyrene beads and microcrystals (ferrocene microcrystal) can be utilized for the highly sensitive electrochemical immunoassays. Polystyrene beads coated with nucleic acid were utilized for an amplified electrochemical sandwich immunoassay of

proteins. Similarly, different oligonucleotide tracers utilizing guanine to adenine ratio provide a good alternative for multiplex protein assay. Recently, NP encoded spheres, submicrometer metallic microrods, metal or alloy nanowires were utilized for product identification, electrochemical DNA hybridization assay, and optical bioassay. Semiconductor NP encoded polystyrene beads and metallic striped nanowires can be used as tags for multiplex optical immunoassay of DNA, proteins, and pathogens. **Nanowires** are metallic and nonmetallic 1D-nanostructure. Nanowire-conjugated enzymes are explored for developing biosensing devices and electrocatalytic applications due to their high aspect ratio, high surface-to-volume ratio, low or no toxicity, and biocompatibility (Bagal-Kestwal et al., 2018; Liu and Lin, 2007).

2.15. Upconverting nanoparticles (UCNPs)

Upconverting nanoparticles (UCNPs) absorb photons in the IR range and emit them in the visible range via a nonlinear optical process. UNCPs can replace fluorescent labels for biosensing applications due to low background emission, high excitation wavelength, and lack of absorption effects with other components of the biosensor. Another important application is the transduction that involves nonradiative energy transfer from excited UNCPs to an acceptor leading to luminescence nonradiative energy transfer (LRET) (Wang et al., 2011). Wang et al. (2005) have reported LRET between biotinylated-UNCPs and biotinylated-AuNPs to detect avidin; attached thrombin-aptamer with UNCP to detect thrombin in human plasma or modified UNCP with concanavalin A (Wang et al., 2011; Zhang et al., 2011). The use of UNCP for biosensing has one limitation, i.e., the requirement of 980 nm of wavelength, which lies in the absorption range of water. This might raise the temperature of the sample thus interfering with the bio-imaging process. However, the use of neodymium lowers the excitation wavelength to 808 nm (Himmelstoß et al., 2017; Wang et al., 2013).

2.16. Magnetic beads (MBs)

MBs provide a large surface area for immobilization and solid support for the isolation, purification, and pre-concentration of the analytes of interest thus significantly improving the electrochemical biosensing performance in terms of sensitivity, reproducibility, and reduced assay turnaround. Being dispersible in the solution, MBs enhances the capture efficiency of cancer biomarkers. MBs enhance the selectivity and specificity by carrying out the biorecognition and transduction steps on different surfaces thus reducing the chance of nonspecific adsorptions. SPEs are used for the development of portable, low sample consumption electrochemical biosensors. The planar surface of SPEs aid in their modifications and linkage with MBs via a magnetic field (Cetó et al., 2012). A disposable SPE based on "biomimetic" (oxidase-like activity) cerium (IV) oxide NPs (nanoceria) was used for the detection of wine phenolic antioxidants. Nanoceria-enabled sensors facilitate oxidation of PCs to corresponding quinones that are detected at the SPE surface.

2.17. Conclusion and future outlook

Rapid and progressive development of nanotechnology has ensured the design and fabrication of novel electrochemical biosensors with improvised features. The arrival of nanomaterial-based labels for electrochemical biosensing applications has facilitated the design of electrochemical biosensors/immunosensors with improvised features. Nanomaterial-based electrochemical biosensing leads to enhanced transducing signals of ligand-analyte interaction in addition to high selectivity, specificity, lower detection limit, multiplexing ability, miniaturization, customization, low-sample requirement, and lower cost. These features put them at a distinct advantage over conventional approaches for the detection of clinically relevant biomarkers for disease diagnosis, food-adulterants, and environmental monitoring. However, challenges remain in terms of validation of laboratory results with real-life clinical samples; applicability of biosensing protocol over a range of biomarkers for a particular disease; and precise detection of biomarkers when analytes are available in femto- to picomolar concentration range in the huge background of nontarget species. There is a need to develop inexpensive, reproducible, electrochemically favorable, and chemically stable nanomaterials that facilitate easy sample preparation, is free from matrix effects, have long-term storage, stability, multiplexing ability, and integration with portable and automated systems, before the nanomaterial-based biosensing system could achieve its full potential in health care needs, environmental monitoring, and maintenance of food quality. More research is needed from people of diverse backgrounds to design a nanomaterial-based biosensor with universal applicability.

References

Adhikari, B.R., Govindhan, M., Chen, A., 2015. Carbon nanomaterials based electrochemical sensors/biosensors for the sensitive detection of pharmaceutical and biological compounds. Sensors 15, 22490–22508.

Ahmadalinezhad, A., Wu, G., Chen, A., 2011. Mediator-free electrochemical biosensor based on buckypaper with enhanced stability and sensitivity for glucose detection. Biosens. Bioelectron. 30, 287–293.

Aizawa, M., Morioka, A., Suzuki, S., Nagamura, Y., 1979. Enzyme immunosensor. III. Amperometric determination of human chorionic gonadotropin by membrane-bound antibody. Anal. Biochem. 94, 22–28.

Algar, W.R., Khachatrian, A., Melinger, J.S., Huston, A.L., Stewart, M.H., Susumu, K., Blanco-Canosa, J.B., Oh, E., Dawson, P.E., Medintz, I.L., 2017. Concurrent modulation of quantum dot photoluminescence using a combination of charge transfer and Förster resonance energy transfer: competitive quenching and multiplexed biosensing modality. J. Am. Chem. Soc. 139, 363–372.

Ambrosi, A., Castaneda, M.T., Killard, A.J., Smyth, M.R., Alegret, S., Merkoci, A., 2007. Double-codified gold nanolabels for enhanced immunoanalysis. Anal. Chem. 79, 5232.

Anker, J.N., Hall, W.P., Lyandres, O., Shah, N.C., Zhao, J., Van Duyne, R.P., 2008. Biosensing with plasmonic nanosensors. Nat. Mater. 7, 442–453.

Bagal-Kestwal, D.R., Pan, M.H., Chiang, B.H., 2018. Electrically nanowired-enzymes for probe modification and sensor fabrication. Biosens. Bioelectron. 121, 223–235.

Bai, Y., Feng, F., Zhao, L., Wang, C., Wang, H., Tian, M., Qin, J., Duan, Y., He, X., 2013. Aptamer/thrombin/aptamer-AuNPs sandwich enhanced surface plasmon resonance sensor for the detection of subnanomolar thrombin. Biosens. Bioelectron. 47, 265–270.

Bakker, E., Qin, Y., 2006. Electrochemical sensors. Anal. Chem. 78, 3965.

Barabadi, H., Najafi, M., Samadian, H., Azarnezhad, A., Vahidi, H., Mahjoub, M.A., Koohiyan, M., Ahmadi, A., 2019. A systematic review of the genotoxicity and antigenotoxicity of biologically synthesized metallic nanomaterials: are green nanoparticles safe enough for clinical marketing? Medicina (Kaunas) 55, 439.

Bergwerff, A.A., van Knapen, F., 2006. Surface plasmon resonance biosensors for detection of pathogenic microorganisms: strategies to secure food and environmental safety. J. AOAC Int. 89, 826–831.

Blasco, A.J., González Crevillén, A., González, M.C., Escarpa, A., 2007. Direct electrochemical sensing and detection of natural antioxidants and antioxidant capacity in vitro systems. Electroanalysis 19, 2275–2286.

Bollella, P., Fusco, G., Tortolini, C., Sanzò, G., Favero, G., Gorton, L., Antiochia, R., 2017. Beyond graphene: electrochemical sensors and biosensors for biomarkers detection. Biosens. Bioelectron. 89, 152–166.

Bu, D., Zhuang, H., Yang, G., Ping, X., 2014. An immunosensor designed for polybrominated biphenyl detection based on fluorescence resonance energy transfer (FRET) between carbon dots and gold nanoparticles. Sens. Actuators B Chem. 195, 540–548.

Campuzano, S., Pedrero, M., Nikoleli, G.P., Pingarrón, J.M., Nikolelis, D.P., 2017a. Hybrid 2D-nanomaterials-based electrochemical immunosensing strategies for clinical biomarkers determination. Biosens. Bioelectron. 89, 269–279.

Campuzano, S., Pedrero, M., Pingarrón, J.M., 2017b. Non-invasive breast cancer diagnosis through electrochemical biosensing at different molecular levels. Sensors 17, 1993.

Campuzano, S., Yáñez-Sedeño, P., Pingarrón, J.M., 2017c. Electrochemical Genosensing of circulating biomarkers. Sensors 17, 866.

Campuzano, S., Yáñez-Sedeño, P., Pingarrón, J.M., 2017d. Molecular biosensors for electrochemical detection of infectious pathogens in liquid biopsies: current trends and challenges. Sensors 17, 2533.

Cao, J., Sun, T., Grattan, K.T.V., 2014. Gold nanorod-based localized surface plasmon resonance biosensors: a review. Sens. Actuators B Chem. 195, 332–351.

Cardoso, C.E.D., Almeida, J.C., Lopes, C.B., Trindade, T., Vale, C., Pereira, E., 2019. Recovery of rare earth elements by carbon-based nanomaterials—a review. Nanomaterials (Basel) 9, 814.

Cetó, X., Céspedes, F., Pividori, M.I., Gutiérrez, J.M., del Valle, M., 2012. Resolution of phenolic antioxidant mixtures employing a voltammetric bio-electronic tongue. Analyst 137, 349–356.

Chen, X., Wang, Y., Zhou, J., Yan, W., Li, X., Zhu, J.J., 2008. Electrochemical impedance immunosensor based on three-dimensionally ordered macroporous gold film. Anal. Chem. 80, 2133–2140.

Chikkaveeraiah, B.V., Mania, V., Patel, V., Gutkind, J.S., Rusling, J.F., 2011. Microfluidic electrochemical immunoarray for ultrasensitive detection of two cancer biomarker proteins in serum. Biosens. Bioelectron. 26, 4477–4483.

Chouhan, R.S., Žitko, G., Fajon, V., Živković, I., Pavlin, M., Berisha, S., Jerman, I., Vesel, A., Horvat, M., 2019. A unique interactive nanostructure knitting based passive sampler adsorbent for monitoring of Hg2+ in water. Sensors (Basel) 19, 3432.

Chu, X., Zhao, Z., Shen, G., Yu, R., 2006. Quartz crystal microbalance immunoassay with dendritic amplification using colloidal gold immunocomplex. Sens. Actuators B 114, 696.

Chumbimuni-Torres, K., Dai, Z., Rubinova, N., Xiang, Y., Pretsch, E., Wang, J., Bakker, E., 2006. Potentiometric biosensing of proteins with ultrasensitive ion-selective microelectrodes and nanoparticle labels. J. Am. Chem. Soc. 128, 13676.

Cui, Z., Wu, D., Zhang, Y., Ma, H., Li, H., Du, B., Wei, Q., Ju, H., 2014. Ultrasensitive electrochemical immunosensors for multiplexed determination using mesoporous platinum nanoparticles as nonenzymatic labels. Anal. Chim. Acta 807, 44–50.

Darwish, N.T., Alrawi, A.H., Sekaran, S.D., Alias, Y., Khor, S.M., 2016. Electrochemical immunosensor based on antibody-nanoparticle hybrid for specific detection of the dengue virus NS1 biomarker. J. Electrochem. Soc. 63, B19–B25.

Dequaire, M., Degrand, C., Limoges, B., 2000. An electrochemical metalloimmunoassay based on a colloidal gold label. Anal. Chem. 72, 5521.

Ding, Y.Y., Li, D., Li, B., Du, W., Zheng, J.Y., Yang, M.H., 2013. A water-dispersible, ferrocene-tagged peptide nanowire for amplified electrochemical immunosensing. Biosens. Bioelectron. 48, 281–286.

Doria, G., Conde, J., Veigas, B., Giestas, L., Almeida, C., Assunção, M., Rosa, J., Baptista, P.V., 2012. Noble metal nanoparticles for biosensing applications. Sensors 12, 1657–1687.

El-Ansary, A., Faddah, L.M., 2010. Nanoparticles as biochemical sensors. Nanotechnol. Sci. Appl. 3, 65–76.

Esteban-Fernández de Ávila, B., Araque, E., Campuzano, S., Pedrero, M., Dalkiran, B., Barderas, R., Villalonga, R., Kiliç, E., Pingarrón, J.M., 2015. Dual functional graphene derivative-based electrochemical platforms for detection of the TP53 gene with single nucleotide polymorphism selectivity in biological samples. Anal. Chem. 87, 2290–2298.

Fathi, F., Rashidi, M.R., Omidi, Y., 2019. Ultra-sensitive detection by metal nanoparticles-mediated enhanced SPR biosensors. Talanta 192, 118–127.

Felix, F.S., Angnes, L., 2018. Electrochemical immunosensors—a powerful tool for analytical applications. Biosens. Bioelectron. 15 (102), 470–478.

Fu, Y.Z., Yuan, R., Tang, D.P., Chai, Y.Q., Xu, L., 2005. Study on the immobilization of anti-IgG on Au-colloid modified gold electrode via potentiometric immunosensor, cyclic voltammetry, and electrochemical impedance techniques. Colloids Surf. B. Biointerfaces 40, 61–66.

Garcia, M.F.K., Andrade, C.A.S., de Melo, C.P., Gomes, D.S., Silva, L.G., Dias, R.V., Balbino, V.Q., Oliveira, M.D.L., 2016. Impedimetric sensor for *Leishmania infantum* genome based on gold nanoparticles dispersed in polyaniline matrix. J. Chem. Technol. Biotechnol. 91, 2810–2816.

Gayathri, C.H., Mayuri, P., Sankaran, K., Kumar, A.S., 2016. An electrochemical immunosensor for efficient detection of uropathogenic *E. coli* based on thionine dye immobilized chitosan/functionalized-MWCNT modified electrode. Biosens. Bioelectron. 82, 71–77.

Ghalkhani, M., Ghorbani-Bidkorbeh, F., 2019. Development of carbon nanostructured based electrochemical sensors for pharmaceutical analysis. Iran. J. Pharm. Res. 18, 658–669.

Ghindilis, L., Atanasov, P., Wilkinst, M., Wilkins, E., 1998. Immunosensors: electrochemical sensing and other engineering approaches. Biosens. Bioelectron. 13, 113.

Ghindilis, A.L., Skorobotko, O.V., Gavnlova, V.P., Yaropolov, A.I., 1992. A new approach to the construction of potentiometric immunosensors. Biosens. Bioelectron. 7, 301–304.

Gonzalez-Garcia, M.B., Fernandez-Sanchez, C., Costa-Garcia, A., 2000. Colloidal gold as an electrochemical label of streptavidin-biotin interaction. Biosens. Bioelectron. 15, 315.

Haque, E., Anwar Hossain, M., Namihira, Y., Ahmed, F., 2019. Microchannel-based plasmonic refractive index sensor for low refractive index detection. Appl. Opt. 58, 1547–1554.

Hassani, H., Kreysing, E., 2019. Noninvasive measurement of the refractive index of cell organelles using surface plasmon resonance microscopy. Opt. Lett. 44, 1359–1362.

Hayatt, M.A., 1989. Colloidal Gold-Principles, Methods and Applications. Academic Press, San Diego.

Hemmateenejad, B., Shamsipur, M., Khosousi, T., Shanehsaz, M., Firuzi, O., 2012. Antioxidant activity assay based on the inhibition of oxidation and photobleaching of L-cysteine-capped CdTe quantum dots. Analyst 137, 4029.

Himmelstoß, S.F., Wiesholler, L.M., Buchner, M., Muhr, V., Märkl, S., Baeumner, A.J., Hirsch, T., 2017. 980 nm and 808 nm excitable upconversion nanoparticles for the detection of enzyme related reactions. Proc. SPIE, 10077.

Holzinger, M., Goff, A.L., Cosnier, S., 2017. Synergetic effects of combined nanomaterials for biosensing applications. Sensors 17, 1010.

Hong, C.L., Yuan, R., Chai, Y.Q., Zhuo, Y., 2009. Ferrocenyl-doped silica nanoparticles as an immobilized affinity support for electrochemical immunoassay of cancer antigen 15-3. Anal. Chim. Acta 633, 244–249.

Hou, Y.H., Wang, J.J., Jiang, Y.Z., Lv, C., Xia, L., Hong, S.L., Lin, M., Lin, Y., Zhang, Z.L., Pang, D.W., 2017. A colorimetric and electrochemical immunosensor for point-of-care detection of enterovirus 71. Biosens. Bioelectron. 99, 186–192.

Jeon, J., 2019. Review of therapeutic applications of radiolabeled functional nanomaterials. Int. J. Mol. Sci. 20, 2323.

Jeong, B.J., Akter, R., Han, O.H., Rhee, C.K., Rahman, M.A., 2013. Increased electrocatalyzed performance through dendrimer-encapsulated gold nanoparticles and carbon nanotube-assisted multiple bienzymatic labels: highly sensitive electrochemical immunosensor for protein detection. Anal. Chem. 85, 1784–1791.

Jiao, L., Mu, Z.G., Zhu, C.Z., Wei, Q., Li, H., Du, D., Lin, Y.H., 2016. Graphene loaded bimetallic Au@Pt nanodendrites enhancing ultrasensitive electrochemical immunoassay of AFP. Sens. Actuators B Chem. 231, 513–519.

Kharitonov, A.B., Alfonta, L., Katz, E., Willner, I., 2000. Probing of bioaffinity interactions at interfaces using impedance spectroscopy and chronopotentiometry. J. Electroanal. Chem. 487, 133–141.

Kuo, W.K., Tongpakpanang, J., Kuo, P.H., Kuo, S.F., 2019. Implementation and phase detection of dielectric-grating-coupled surface plasmon resonance sensor for backside incident light. Opt. Express 27, 3867–3872.

Li, L., Feng, D., Zhang, Y., 2016a. Simultaneous detection of two tumor markers using silver and gold nanoparticles decorated carbon nanospheres as labels. Anal. Biochem. 505, 59–65.

Li, J., Skeete, Z., Shan, S., Yan, S., Kurzatkowska, K., Zhao, W., Ngo, Q.M., Holubovska, P., Luo, J., Hepel, M., Zhong, C.J., 2015. Surface enhanced Raman scattering detection of cancer biomarkers with bifunctional nanocomposite probes. Anal. Chem. 87, 10696–10702.

Li, X., Wei, J., Aifantis, K.E., Fan, Y., Feng, Q., Cui, F.Z., Watari, F., 2016b. Current investigations into magnetic nanoparticles for biomedical applications. J. Biomed. Mater. Res. A 104, 1285–1296.

Liao, H., Nehl, C.L., Hafner, J.H., 2006. Biomedical applications of plasmon resonant metal nanoparticles. Nanomedicine 1, 201–208.

Lin, J., Ju, H., 2005. Electrochemical and chemiluminescent immunosensors for tumor markers. Biosens. Bioelectron. 20, 1461.

Liu, G., Lin, Y., 2007. Nanomaterial labels in electrochemical immunosensors and immunoassays. Talanta 74, 308.

Liu, G., Wang, J., Wu, H., Lin, Y., Lin, Y., 2007. Nanovehicles based bioassay labels. Electroanalysis 19, 777.

Liu, X.Q., Wang, X.H., Zhang, J.M., Feng, X.H., Liu, X.H., 2012. Detection of estradiol at an electrochemical immunosensor with a Cu UPD|DTBP-protein G scaffold. Biosens. Bioelectron. 35, 56–62.

Liu, B.Q., Zhang, B., Cui, Y.L., 2011. Multifunctional gold-silica nanostructures for ultrasensitive electrochemical immunoassay of streptomycin residues. ACS Appl. Mater. Interfaces 3, 4668–4676.

Liu, Y., Zhao, Y., Zhang, T., Chang, Y., Wang, S., Zou, R., Zhu, G., Shen, L., Guo, Y., 2019. Quantum dots-based immunochromatographic strip for rapid and sensitive detection of acetamiprid in agricultural products. Front. Chem. 7, 76.

Luna, D.M.N., Avelino, K.Y.P.S., Cordeiro, M.T., Andrade, C.A.S., Oliveira, M.D.L., 2015. Electrochemical immunosensor for dengue virus serotypes based on 4-mercaptobenzoic acid modified gold nanoparticles on self-assembled cysteine monolayers. Sens. Actuators B Chem. 220, 565–572.

Ma, T.Y., Tang, Y., Dai, S., Qiao, S.Z., 2014. Proton-functionalized two-dimensional graphitic carbon nitride nanosheet: an excellent metal-/label-free biosensing platform. Small 10, 2382–2389.

Maduraiveeran, G., Sasidharan, M., Ganesan, V., 2018. Electrochemical sensor and biosensor platforms based on advanced nanomaterials for biological and biomedical applications. Biosens. Bioelectron. 103, 113–129.

Malic, L., Sandros, M.G., Tabrizian, M., 2011. Designed biointerface using near-infrared quantum dots for ultrasensitive surface plasmon resonance imaging biosensors. Anal. Chem. 83, 5222–5229.

Manikandan, V.S., Adhikari, B., Chen, A., 2018. Nanomaterial based electrochemical sensors for the safety and quality control of food and beverages. Analyst 143, 4537–4554.

Mao, X., Jiang, J.H., Chen, J.W., Huang, Y., Shen, G.L., Yu, R.Q., 2006. Cyclic accumulation of nanoparticles: a new strategy for electrochemical immunoassay based on the reversible reaction between dethiobiotin and avidin. Anal. Chim. Acta 557, 159.

Marquette, C.A., Blum, L.J., 2006. State of the art and recent advances in immunoanalytical systems. Biosens. Bioelectron. 21, 1424.

Mirsky, V.M., Riepl, M., Wolfbeis, O.S., 1997. Capacitive monitoring of protein immobilization and antigen-antibody reactions on monomolecularalkylthiol films on gold electrodes. Biosens. Bioelectron. 12, 977–989.

Mohammadzadeh-Asl, S., Keshtkar, A., Dolatabadi, E.N., de la Guardia, M., 2018. Nanomaterials and phase sensitive based signal enhancement in surface plasmon resonance. Biosens. Bioelectron. 110, 118–131.

Moon, S.W., Ha, J.W., 2019. Single-particle correlation study: chemical interface damping induced by biotinylated proteins with sulfur in plasmonic gold nanorods. Phys. Chem. Chem. Phys. 21, 7061–7066.

Moro, L., Turemis, M., Marini, B., Ippodrino, R., Giardi, M.T., 2017. Better together: strategies based on magnetic particles and quantum dots for improved biosensing. Biotechnol. Adv. 35, 51–63.

Muhammad-Tahir, Z., Alocilja, E.C., 2003. A conductometric biosensor for biosecurity. Biosens. Bioelectron. 18, 813–819.

Navya, P.N., Kaphle, A., Srinivas, S.P., Bhargava, S.K., Rotello, V.M., Daima, H.K., 2019. Current trends and challenges in cancer management and therapy using designer nanomaterials. Nano Converg. 6, 23.

Pan, M., Yagauang, Y.G., Li, M., Zing, X., Wang, S., 2017. Nanomaterials for electrochemical immunosensing. Sensors 17, 1041.

Peng, D., Liang, R.P., Huang, H., Qiu, J.D., 2016. Electrochemical immunosensor for carcinoembryonic antigen based on signal amplification strategy of graphene and Fe_3O_4/au NPs. J. Electroanal. Chem. 761, 112–117.

Peng, M.P., Ma, W., Long, Y.T., 2015. Alcohol dehydrogenase-catalyzed gold nanoparticle seed-mediated growth allows reliable detection of disease biomarkers with the naked eye. Anal. Chem. 87, 5891–5896.

Qiu, H.J., Guan, Y., Luo, P., Wang, Y., 2017. Recent advance in fabricating monolithic 3D porous graphene and their applications in biosensing and biofuel cells. Biosens. Bioelectron. 89, 85–95.

Sánchez, J.A.L., Díez-Pascual, A.M., Capilla, R.P., Díaz, R.P., 2019. The effect of hexamethylene diisocyanate-modified graphene oxide as a nanofiller material on the properties of conductive polyaniline. Polymers (Basel) 11, 1032.

Sánchez-Tirado, R., Salvo, C., González-Cortés, A., Yáñez-Sedeño, P., Langa, F., Pingarrón, J.M., 2017. Electrochemical immunosensor for simultaneous determination of interleukin-1 beta and tumor necrosis factor alpha in serum and saliva using dual screen printed electrodes modified with functionalized double-walled carbon nanotubes. Anal. Chim. Acta 959, 66–73.

Shuai, H.L., Huang, K.J., Xing, L.L., Chen, Y.X., 2016. Ultrasensitive electrochemical sensing platform for microRNA based on tungsten oxide-graphene composites coupling with catalyzed hairpin assembly target recycling and enzyme signal amplification. Biosens. Bioelectron. 86, 337–345.

Singh, P., 2014. Surface Plasmon Resonance. Nova Science Publishers, New York, USA.

Singh, P., 2016a. SPR biosensors: historical perspectives and current challenges. Sensors Actuators B Chem. 229, 110–130.

Singh, P., 2016b. In: Giddes, C.D. (Ed.), LSPR Biosensing: Recent Advances and Approaches in Reviews in Plasmonics. vol. 2017. Springer International Publishing, AG, pp. 211–238.

Singh, P., 2017a. G protein coupled receptors: a potential candidate for drug discovery. Int. J. Pharmacol. 13, 846–863 (Special Issue).

Singh, P., 2017b. Surface Plasmon Resonance: A Boon for Viral Diagnostics in Reference Module in Life Sciences. Elsevier, Netherlands.

Singh, M., Holzinger, M., Tabrizian, M., Winters, S., Berner, N.C., Cosnier, S., Duesberg, G.S., 2015. Noncovalently functionalized monolayer graphene for sensitivity enhancement of surface plasmon resonance immunosensors. J. Am. Chem. Soc. 137, 2800–2803.

Stradiotto, N.R., Yamanaka, H., Zanoni, M.V.B., 2003. Electrochemical sensors: a powerful tool in analytical chemistry. J. Braz. Chem. Soc. 14, 159.

Sun, G., Zhang, L., Zhang, Y., Yang, H., Ma, C., Ge, S., Yan, M., Yu, J., Song, X., 2015. Multiplexed enzyme-free electrochemical immunosensor based on ZnO nanorods modified reduced graphene oxide-paper electrode and silver deposition-induced signal amplification strategy. Biosens. Bioelectron. 71, 30–36.

Tang, J., Huang, J.X., Su, B.L., Chen, H.F., Tang, D.P., 2011a. Sandwich-type conductometric immunoassay of alpha-fetoprotein in human serum using carbon nanoparticles as labels. Biochem. Eng. J. 53, 223–228.

Tang, D.P., Su, B.L., Tang, J., Ren, J.J., Chen, G.N., 2010. Nanoparticle-based sandwich electrochemical immunoassay for carbohydrate antigen 125 with signal enhancement using enzyme-coated nanometer-sized enzyme-doped silica beads. Anal. Chem. 82, 1527–1534.

Tang, J., Tang, D.R., Su, B.L., Huang, J.X., Qiu, B., Chen, G.N., 2011b. Enzyme-free electrochemical immunoassay with catalytic reduction of p-nitrophenol and recycling of p-aminophenol using gold nanoparticles-coated carbon nanotubes as nanocatalysts. Biosens. Bioelectron. 7, 3219–3226.

Tang, D.P., Yuan, R., Chai, Y.Q., Liu, Y., Zhong, X., Dai, J.Y., 2004. Potentiometric diphtheria immunosensor based on a glassy carbon electrode modified with colloidal gold and anti-diph. Acta Chim. Sin. 62, 2062–2066.

Torrente-Rodríguez, R.M., Ruiz-Valdepeñas, V., Campuzano, M.S., Farchado, P.M., Vargas, E., de Villena, F.J.M., Garranzo-Asensio, M., Barderas, R., Pingarrón, J.M., 2017. Electrochemical sensor for rapid determination of fibroblast growth factor receptor 4 in raw cancer cell lysates. PLoS One 12, e0175056.

Velev, O.D., Kaler, E.W., 1999. *In Situ* assembly of colloidal particles into miniaturized biosensors. Langmuir 15, 3693.

Verma, P., Maire, P., Novak, P., 2011. Concatenation of electrochemical grafting with chemical or electrochemical modification for preparing electrodes with specific surface functionality. Electrochim. Acta 56, 3555–3561.

Wanekaya, A.K., 2011. Applications of nanoscale carbon-based materials in heavy metal sensing and detection. Analyst 136, 4383–4391.

Wang, J., 2007. Nanopraticle based electrochemical bioassays of proteins. Electroanalysis 19, 769.

Wang, F., Deng, R., Wang, J., Wang, Q., Han, Y., Zhu, H., Chen, X., Liu, X., 2011. Tuning upconversion through energy migration in core-shell nanoparticles. Nat. Mater. 10, 968–973.

Wang, X., Ge, L., Yu, Y., Dong, S., Li, F., 2015. Highly sensitive electrogenerated chemiluminescence biosensor based on hybridization chain reaction and amplification of gold nanoparticles for DNA detection. Sens. Actuator B 220. 942–98.

Wang, K., He, M.Q., Zhai, F.H., He, R.H., Yu, Y.L., 2017. A novel electrochemical biosensor based on polyadenine modified aptamer for label-free and ultrasensitive detection of human breast cancer cells. Talanta 166, 87–92.

Wang, J., Liu, G., Engelhard, M.H., Lin, Y., 2006. Sensitive immunoassay of a biomarker tumor necrosis factor-alpha based on poly(guanine)-functionalized silica nanoparticle label. Anal. Chem. 78, 6974.

Wang, J., Liu, G.D., Jan, M.R., 2004. Ultrasensitive electrical biosensing of proteins and DNA: carbon-nanotube derived amplification of the recognition and transduction events. J. Am. Chem. Soc. 126, 3010–3011.

Wang, Y.F., Liu, G.Y., Sun, L.D., Xiao, J.W., Zhou, J.C., Yan, C.H., 2013. Nd(3+)-sensitized upconversion nanophosphors: efficient in vivo bioimaging probes with minimized heating effect. ACS Nano 7, 7200–7206.

Wang, J., Munir, A., Zhu, Z., Zhou, H.S., 2010. Magnetic nanoparticle enhanced surface plasmon resonance sensing and its application for the ultrasensitive detection of magnetic nanoparticle-enriched small molecules. Anal. Chem. 82, 6782–6789.

Wang, Y., Partridge, A., Wu, Y., 2019. Improving nanoparticle-enhanced surface plasmon resonance detection of small molecules by reducing steric hindrance via molecular linkers. Talanta 1 (198), 350–357.

Wang, L., Yan, R., Huo, Z., Wang, L., Zeng, J., Bao, J., Wang, X., Peng, Q., Li, Y., 2005. Fluorescence resonant energy transfer biosensor based on upconversion-luminescent nanoparticles. Angew. Chem. Int. Ed. Engl. 44, 6054–6057.

Warsinke, A., Benkert, A., Scheller, F.W., 2000. Electrochemical immunoassays. Fresenius J. Anal. Chem. 366, 622.

Wu, Y.F., Chen, C.L., Liu, S.Q., 2009. Enzyme-functionalized silica nanoparticles as sensitive labels in biosensing. Anal. Chem. 81, 1600–1607.

Wu, L., Ji, H., Sun, H., Ding, C., Ren, J., Qu, X., 2016b. Label-free ratiometric electrochemical detection of the mutated apolipoprotein E gene associated with Alzheimer's disease. Chem. Commun. 52, 12080–12083.

Wu, D., Li, R., Wang, H.X., Liu, S., Wang, H., Wei, Q., Du, B., 2012. Hollow mesoporous silica microspheres as sensitive labels for immunoassay of prostate-specific antigen. Analyst 137, 608–613.

Wu, D., Liu, Y., Wang, Y., Hu, L., Ma, H., Wang, G., Wei, Q., 2016a. Label-free electrochemiluminescent immunosensor for detection of prostate specific antigen based on aminated graphene quantum dots and carboxyl graphene quantum dots. Sci. Rep. 6, 20511.

Xiong, M., Rong, Q., Meng, H.M., Zhang, X.B., 2017. Two-dimensional graphitic carbon nitride nanosheets for biosensing applications. Biosens. Bioelectron. 89, 212–223.

Xu, T.S., 2016. Bioconjugation of peroxidase-like nanostructures with natural enzyme for in-situ amplified conductometric immunoassay of tissue polypeptide antigen in biological fluids. Biochem. Eng. J. 105, 36–43.

Yagiuda, K., Hemmi, A., Ito, S., Asano, Y., 1996. Development of a conductivity-based immunosensor for sensitive detection of methamphetamine (stimulant drug) in human urine. Biosens. Bioelectron. 11, 703–707.

Yáñez-Sedeño, P., Campuzano, S., Pingarrón, J.M., 2017. Multiplexed electrochemical immunosensors for clinical biomarkers. Sensors 17, 965.

Yáñez-Sedeño, P., Campuzano, S., Pingarrón, J.M., 2019. Pushing the limits of electrochemistry toward challenging applications in clinical diagnosis, prognosis, and therapeutic action. Chem. Commun. 55 (18), 2563–2592.

Yang, M.H., Li, H., Javadi, A., Gong, S.Q., 2010. Multifunctional mesoporous silica nanoparticles as labels for the preparation of ultrasensitive electrochemical immunosensors. Biomaterials 31, 3281–3286.

Yang, Z.H., Zhuo, Y., Chai, Y.Q., Yuan, R., 2014. High throughput immunosenor based on multi-label strategy and a novel array electrode. Sci. Rep. 4, 4747.

Yu, X., Munge, B., Patel, V., Jensen, G., Bhirde, A., Gong, J.D., Kim, S.N., Gillespie, J., Gutkind, J.S., Papadimitrakopoulos, F., Rusling, J.F., 2006. Carbon nanotube amplification strategies for highly sensitive immunodetection of cancer biomarkers. J. Am. Chem. Soc. 128, 11199–11205.

Yu, X., Zhang, W., Zhang, P., Su, Z., 2017. Fabrication technologies and sensing applications of graphene-based composite films: advances and challenges. Biosens. Bioelectron. 89, 72–84.

Yuan, J., Oliver, R., Li, J., Lee, J., Aguilar, M., Wu, Y., 2007. Sensitivity enhancement of SPR assay of progesterone based on mixed self-assembled monolayers using nanogold particles. Biosens. Bioelectron. 23 (1), 144–148.

Zhang, G.J., Luo, Z.H.H., Huang, M.J., Ang, J.J., Kang, T.G., Ji, H., 2011. An integrated chip for rapid, sensitive, and multiplexed detection of cardiac biomarkers from fingerprick blood. Biosens. Bioelectron. 28, 459–463.

Zhang, S., Ma, H.M., Yan, L.G., Cao, W., Yan, T., Wei, Q., Du, B., 2014. Copper-doped titanium dioxide nanoparticles as dual-functional labels for fabrication of electrochemical immunosensors. Biosens. Bioelectron. 59, 335–341.

Zheng, L., Jia, L., Li, B., Situ, B., Liu, Q., Wang, Q., Gan, N., 2012. A sandwich HIV p24 amperometric immunosensor based on a direct gold electroplating-modified electrode. Molecules 17, 5988–6000.

Zheng, L., Niu, X., Zhao, J., Liu, T., Liu, Y., Yang, Y., 2016. A label-free immunosensor for CEA based on Pd-Ir bimetallic nanoparticles. J. Nanosci. Nanotechnol. 16, 5984–5990.

Zhou, W.J., Halpern, A.R., Seefeld, T.H., Corn, R.M., 2012. Near infrared surface plasmon resonance phase imaging and nanoparticle-enhanced surface plasmon resonance phase imaging for ultrasensitive protein and DNA biosensing with oligonucleotide and aptamer microarrays. Anal. Chem. 84, 440–445.

Zhou, S., Wang, Y., Zhu, J.J., 2016. Simultaneous detection of tumor cell apoptosis regulators Bcl-2 and Bax through a dual-signal-marked electrochemical immunosensor. ACS Appl. Mater. Interfaces 8, 7674–7682.

Zhu, Q., Chai, Y., Zhuo, Y., Yuan, R., 2015. Ultrasensitive simultaneous detection of four biomarkers based on hybridization chain reaction and biotin-streptavidin signal amplification strategy. Biosens. Bioelectron. 68, 42–48.

Zhu, N., Chang, Z., He, P., Fang, Y., 2005. Electrochemical DNA biosensors based on platinum NPs combined carbon nanotubes. Anal. Chim. Acta 545, 21–26.

Zhuo, Y., Yuan, R., Chai, Y.Q., Hong, C.L., 2010. Functionalized SiO2 labeled CA19–9 antibodies: a new strategy for signal amplification of antigen-antibody sensing processes. Analyst 135, 2036–2042.

Ziyatdinova, G., Galandova, J., Labuda, J., 2008. Impedimetric nanostructured disposable DNA-based biosensors for the detection of deep DNA damage and effect of antioxidants. Int. J. Electrochem. Sci. 3, 223–235.

Further reading

Ahmed, A., Rushworth, J.V., Hirst, N.A., Millner, P.A., 2014. Biosensors for whole-cell bacterial detection. Clin. Microbiol. Rev. 27, 631–646.

Benvidi, A., Dehghani Firouzabadi, A., Dehghan Tezerjani, M., Moshtaghiun, S.M., Mazloum-Ardakani, M., Ansarin, A., 2015. A highly sensitive and selective electrochemical DNA biosensor to diagnose breast cancer. J. Electroanal. Chem. 750, 57–64.

Blandón-Naranjo, L., Della Pelle, F., Vázquez, M.V., Gallego, J., Santamaría, A., Alzate-Tobón, M., Compagnone, D., 2018b. Electrochemical behaviour of microwave-assisted oxidized MWCNTs based disposable electrodes: proposal of a NADH electrochemical sensor. Electroanalysis 30, 509–516.

Blandón-Naranjo, L., Hoyos-Arbeláez, J., Vázquez, M.V., Della Pelle, F., Compagnone, D., 2018a. NADH oxidation onto different carbon-based sensors: effect of structure and surface-oxygenated groups. J. Sens. 2018, 6525919.

Bode, A.M., Dong, Z., 2004. Post-translational modification of p53 in tumorigenesis. Nat. Rev. Cancer 4, 793–805.

Castañeda, M.T., Merkoçi, A., Pumera, M., Alegret, S., 2007. Electrochemical genosensors for biomedical applications based on gold nanoparticles. Biosens. Bioelectron. 22, 1961–1967.

Chandra, N., Liu, Y., Liu, J.X., Frängsmyr, L., Wu, N., Silva, L.M., Lindström, M., Chai, W., Pedrosa Domellöf, F., Feizi, T., Arnberg, N., 2019. Sulfated glycosaminoglycans as viral decoy receptors for human adenovirus type 37. Viruses 11 (3). pii: E247.

Chen, L.J., Zeng, G.M., Zhang, Y., Tang, L., Huang, D.L., Liu, C., Pang, Y., Luo, J., 2010. Trace detection of picloram using an electrochemical immunosensor based on three-dimensional gold nanoclusters. Anal. Biochem. 407, 172–179.

Chen, J., Zhang, J., Guo, Y., Li, J., Fu, F., Yang, H.H., Chen, G., 2011. An ultrasensitive electrochemical biosensor for detection of DNA species related to oral cancer based on nuclease-assisted target recycling and amplification of DNAzyme. Chem. Commun. 47, 8004–8006.

Collinge, J., 2011. Prion diseases of humans and animals: their causes and molecular basis. Annu. Rev. Neurosci. 24, 519–550.

Costa Rama, E., Costa-García, A., 2016. Screen-printed electrochemical immunosensors for the detection of cancer and cardiovascular biomarkers. Electroanalysis 28, 1700–1715.

Das, J., Aziz, M.A., Yang, H.A., 2006. A nanocatalyst-based assay for proteins: DNA-free ultrasensitive electrochemical detection using catalytic reduction of p-nitrophenol by gold-nanoparticle labels. J. Am. Chem. Soc. 128, 16022–16023.

Eletxigerra, U., Martinez-Perdiguero, J., Merino, S., Barderas, R., Ruiz-Valdepeñas Montiel, V., Villalonga, R., Pingarrón, J.M., Campuzano, S., 2016. Estrogen receptor α in serum, cell lysates, and breast cancer cell using an amperometric magnetoimmunosensing platform. Sens. Bio Sens. Res. 7, 71–76.

Emami, M., Shamsipur, M., Saber, R., Irajirad, R., 2014. An electrochemical immunosensor for detection of a breast cancer biomarker based on antiHER2-iron oxide nanoparticle bioconjugates. Analyst 139, 2858–2866.

Ensafi, A.A., Nasr-Esfahani, P., Heydari-Bafrooei, E., Rezaei, B., 2014. Redox targeting of DNA anchored to MWCNTs and TiO2 nanoparticles dispersed in poly dialyldimethylammonium chloride and chitosan. Colloids Surf. B Biointerfaces 121, 99–105.

Esteban-Fernández de Ávila, B., Araque, E., Campuzano, S., Pedrero, M., Dalkiran, B., Barderas, R., Villalonga, R., Kiliç, E., Pingarrón, J.M., 2015. Dual functional graphene derivative-based electrochemical platforms for detection of the TP53 gene with single nucleotide polymorphism selectivity in biological samples. Anal. Chem. 87, 2290–2298.

Esteban-Fernández de Ávila, B., Escamilla-Gómez, V., Lanzone, V., Campuzano, S., Pedrero, M., Compagnone, D., Pingarrón, J.M., 2014. Multiplexed determination of amino-terminal pro-B type natriuretic peptide and C-reactive protein cardiac biomarkers in human serum at a disposable electrochemical magnetoimmunosensor. Electroanalysis 26, 254–261.

Farzin, L., Shamsipur, M., Samandari, L., Sheibani, S., 2018. Recent advances in designing nanomaterial based biointerfaces for electrochemical biosensing cardiovascular biomarkers. J. Pharm. Biomed. Anal. 161, 344–376.

Ge, X., Zhang, A., Lin, Y., Du, D., 2016. Simultaneous immunoassay of phosphorylated proteins based on apoferritin templated metallic phosphates as voltammetrically distinguishable signal reporters. Biosens. Bioelectron. 80, 201–207.

Grabowska, I., Sharma, N., Vasilescu, A., Iancu, M., Badea, G., Boukherroub, R., Ogale, S., Szunerits, S., 2018. Electrochemical aptamer-based biosensors for the detection of cardiac biomarkers. ACS Omega 3, 12010–12018.

Gupta, R.K., Pandya, R., Sieffert, T., Meyyappan, M., Koehne, J.E., 2016. Multiplexed electrochemical immunosensor for label-free detection of cardiac markers using a carbon nanofiber array chip. J. Electroanal. Chem. 773, 53–62.

Hernandez, F.J., Ozalp, V.C., 2012. Graphene and other nanomaterial-based electrochemical aptasensors. Biosensors (Basel) 2, 1–14.

Hwang, H., Ryu, M., Park, C.Y., Ahn, J., Park, H.G., Choi, C., Ha, S.D., Park, T.J., Park, J.P., 2017. High sensitive and selective electrochemical biosensor: label-free detection of human norovirus using affinity peptide as molecular binder. Biosens. Bioelectron. 87, 164–170.

Karapetis, S., Nikolelis, D., Hianik, T., 2018. Label-free and redox markers-based electrochemical aptasensors for aflatoxin M1 detection. Sensors (Basel) 18, 4218.

Lin, M., Song, P., Zhou, G., Zuo, X., Aldalbahi, A., Lou, X., Shi, J., Fan, C., 2016. Electrochemical detection of nucleic acids, proteins, small molecules and cells using a DNA-nanostructure-based universal biosensing platform. Nat. Protoc. 11, 1244–1263.

Low, K.F., Rijiravanich, P., Singh, K.K.B., Surareungchai, W., Yean, C.Y., 2015. An electrochemical genosensing assay based on magnetic beads and gold nanoparticle-loaded latex microspheres for vibrio cholerae detection. J. Biomed. Nanotechnol. 11, 702–710.

Mabrouk, M., Rajendran, R., Soliman, I.E., Ashour, M.M., Beherei, H.H., Tohamy, K.M., Thomas, S., Kalarikkal, N., Arthanareeswaran, G., Das, D.B., 2019. Nanoparticle- and nanoporous-membrane-mediated delivery of therapeutics. Pharmaceutics 11, 294.

Manfredi, A., Giannetto, M., Mattarozzi, M., Costantini, M., Mucchino, C., Careri, M., 2016. Competitive immunosensor based on gliadin immobilization on disposable carbon-nanogold screen-printed electrodes for rapid determination of celiotoxic prolamins. Anal. Bioanal. Chem. 408, 7289–7298.

Meisam, O., Malakoutian, M.A., Mohammadmehdi, C., Oroojalian, F., Haghiralsadat, F., Yazdian, F., 2013. A label-free detection of biomolecules using micromechanical biosensors. Chin. Phys. Lett. 30, 068701.

Miodek, A., Castillo, G., Hianik, T., Korri-Youssoufi, H., 2013. Electrochemical aptasensor of human cellular prion based on multiwalled carbon nanotubes modified with dendrimers: a platform for connecting redox markers and aptamers. Anal. Chem. 85, 7704–7712.

Moreno-Guzmán, M., González-Cortés, A., Yáñez-Sedeño, P., Pingarrón, J.M., 2012. Multiplexed ultrasensitive determination of adreno-corticotropinand cortisol hormones at a dual electrochemical immunosensor. Electroanalysis 24, 1100–1108.

Oliveira, N., Souza, E., Ferreira, D., Zanforlin, D., Bezerra, W., Borba, M.A., Arruda, M., Lopes, K., Nascimento, G., Martins, D., Cordeiro, M., Lima-Filho, J., 2015. A sensitive and selective label-free electrochemical DNA biosensor for the detection of specific dengue virus serotype 3 sequences. Sensors 15, 15562–15577.

Pan, Y., Sonn, G.A., Sin, M.L.Y., Mach, K.E., Shih, M.C., Gau, V., Wong, P.K., Liao, J.C., 2010. Electrochemical immunosensor detection of urinary lactoferrin in clinical samples for urinary tract infection diagnosis. Biosens. Bioelectron. 26, 649–654.

Pinto, A.M., Gonçalves, I.C., Magalhães, F.D., 2013. Graphene-based materials biocompatibility: a review. Colloids Surf. B Biointerface 111, 188–202.

Salimian, R., Kékedy-Nagy, L., Ferapontova, E.E., 2017. Specific picomolar detection of a breast cancer biomarker HER-2/neu protein in serum: electrocatalytically amplified electroanalysis by the aptamer/PEG-modified electrode. Chem. Electro. Chem. 4, 872–879.

Shi, H., Wei, J., Qiang, L., Chen, X., Meng, X., 2014. Fluorescent carbon dots for biolmaging and biosensing applications. J. Biomed. Nanotechnol. 10, 2677–2699.

Sun, X.B., Ma, Z.F., 2013. Electrochemical immunosensor based on nanoporpus gold loading thionine for carcinoembryonic antigen. Anal. Chim. Acta 780, 95–100.

Uliana, C.V., Riccardi, C.S., Yamanaka, H., 2014. Diagnostic tests for hepatitis C: recent trends in electrochemical immunosensor and genosensor analysis. World J. Gastroenterol. 20, 15476–15491.

Vasilescu, A., Sharpe, E., Andreescu, S., 2012. Nanoparticle-based technologies for the detection of food antioxidants. Curr. Anal. Chem. 8, 495–505.

Wang, H.J., Bai, L.J., Chai, Y.Q., Yuan, R., 2016. Synthesis of multi-fullerenes encapsulated palladium nanocage, and its application in electrochemiluminescence immunosensors for the detection of Streptococcus suis serotype. Small 2 (10), 1857–1865.

Wang, Y., Dostálek, J., Knoll, W., 2009. Long range surface plasmon-enhanced fluorescence spectroscopy for the detection of aflatoxin M1 in milk. Biosens. Bioelectron. 24, 2264–2267.

Xu, H., Zhu, X., Wang, J., Lin, Z., Chen, G., 2019. Electrochemiluminescent functional nucleic acids-based sensors for food analysis. Luminescence 34, 508–515.

Yadav, S., Nair, S.S., Sai, V.V.R., Satija, J., 2019. Nanomaterials based optical and electrochemical sensing of histamine: progress and perspectives. Food Res. Int. 119, 99–109.

Yang, T., Gao, Y., Liu, Z., Xu, J., Lu, L., Yu, Y., 2017. Three-dimensional gold nanoparticles/prussian blue-poly(3,4-ethylenedioxythiophene) nanocomposite as novel redox matrix for label-free electrochemical immunoassay of carcinoembryonic antigen. Sens. Actuator B Chem. 239, 76–84.

Zhao, C., Liu, X., 2016. A portable paper-based microfluidic platform for multiplexed electrochemical detection of human immunodeficiency virus and hepatitis C virus antibodies in serum. Biomicrofluidics 10, 024119.

Zhao, S., Yang, W., Lai, R.Y., 2011. A folding-based electrochemical aptasensor for detection of vascular endothelial growth factor in human whole blood. Biosens. Bioelectron. 26, 2442–2447.

Zhuo, Z., Ma, M.N., Chai, Y.Q., Zhao, M., Yuan, R., 2014. Amplified electrochemiluminescent aptasensor using mimicking bi-enzyme nanocomplexes as signal enhancement. Anal. Chim. Acta 809, 47–53.

Chapter 3

Electrochemical biosensors: Biomonitoring of clinically significant biomarkers

3.1. Introduction

Human afflictions such as pathogenic infections, various types of cancers, metabolic disorders, genetic abnormalities, and neurological disorders are the major concern for global health. Challenges remain are wrong and delayed diagnosis, which leads to delayed or wrong treatment. Rapid diagnosis of causative factors of the infection plays an important role in controlling disease and optimizing management. Therefore, timely and accurate on-site diagnosis using simple, rapid, and affordable protocols, easily translated to point-of-care-testing (POCT) is crucial for effective treatment and control. Noninvasive diagnosis will particularly lead to effective anti-infective therapy, while simultaneously reducing the quantum of procedure-associated complications (Ahmed et al., 2014). Conventional tests such as cell-culture based or serological tests are slow and time-consuming taking few days to weeks while nucleic acid-based molecular methods such as RT-PCR, ELISA, immunological techniques, etc. are rapid but the cost of infrastructural set-up coupled with the requirement of trained manpower and low sensitivity are big deterrents in resource-limited settings of remote areas, hence not recommended. Electrochemical molecular biosensors address most of these requirements and facilitate simple on-site detection, which is rapid and cost-effective.

This chapter emphasizes the role of electrochemical biosensors in the multiplexed determination of low-abundance critical disease biomarkers in scarcely treated body fluids in a minimally invasive manner for disease diagnosis, prognosis, and point-of-care (POC) applications. This is combined with the development and evolution of electrochemical biosensing platforms in conjunction with nanomaterials for the determination and assessment of other disease biomarkers for metabolic disorders, neurological disorders, and genetic abnormalities. Additionally, the strengths and limitations of the electrochemical sensing platforms and future directions with respect to the emerging, novel pollutants are also discussed. This chapter aims to serve as a complete guide and an important source of information to all the stakeholders including scientists, engineers, and

Electrochemical Biosensors. https://doi.org/10.1016/B978-0-323-90632-6.00004-2

clinicians for augmenting their knowledge on the existing state of electrochemical biosensing for the detection and monitoring of disease biomarkers for diagnostics related point-of-care (POC) applications.

The major aspects of the entire chapter incorporate the following: (i) electrochemical biosensing and its major types based on transducers, (ii) use of different nanomaterials for fabrication and function of electrochemical sensors, (iii) biosensors for the detection of different types of cancers, (iv) biosensors for the monitoring other metabolic, genetic, and neurological disorders, (vi) biosensors for the monitoring pathogens (bacterial and viral), and (vii) recent advances and future trends.

In a typical biosensor, bio-recognition or bioreceptor recognizes target analytes, generating the signal that is transferred to the transducer that performs the detection. The bioreceptors that are most commonly used in biosensors are enzymes, cells, antibodies, molecularly imprinting polymer-(MIP) based molecular recognition, nucleic acids, antibodies, phages, peptides, and aptamers. Other transducers such as electrochemical, impedance, piezoelectric/electrochemical, or voltammetry can be used. Electrochemical biosensors offer inexpensive, field-portable, and programmable battery-operated instrumentation as an exciting alternative for rapid and economical analytical strategies that can be easily handled by unskilled personnel. Affinity based electrochemical biosensors/immunosensors offer a simple, rapid, economical, and minimally invasive protocol for the detection of circulating biomarkers in biofluids such as blood, serum, saliva, urine, sputum, and pleural fluid (for references, Campuzano et al., 2017d) (Fig. 1).

Electrochemical biosensor offers simple, rapid, economical, low-sample requiring, multiplexed, portable, minimal pre-treatment of sample makes them a favorable choice for POC applications (Pan et al., 2010). Nanomaterials such as AuNP, CNT incorporation into sensors offer large surface area, multiple binding sites, conductivity and biocompatibility, sensitivity, selectivity, prolonged stability, and minimum interference from biological matrices hence preferred for the design of affinity-based electrochemical biosensors. These sensors utilize ss-oligonucleotide, peptide sequences, and antibodies. The required sensitivity is achieved by linking nucleic acid-based biosensors with different nucleic acid amplification protocols, which range from PCR to loop-mediated amplification (LAMP), and rolling-circle amplification (RCA), or mimicking in vivo

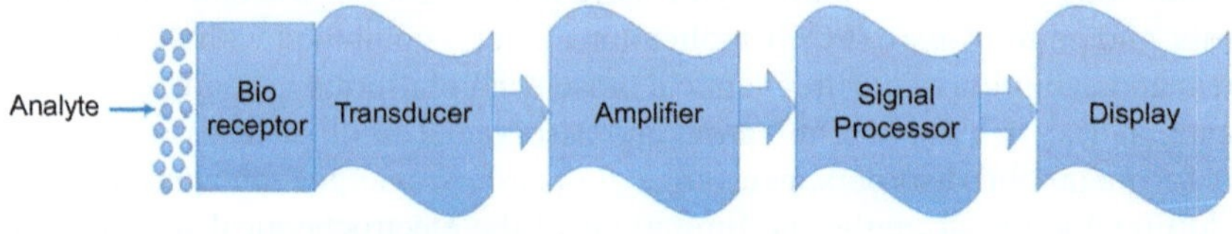

FIG. 1 Schematic diagram of a biosensor (Majdinasab et al., 2017). Copyright © 2017 by the authors. Licensee MDPI, Basel, Switzerland. This article is an open-access article distributed under the terms and conditions of the Creative Commons Attribution (CC BY) license (http://creativecommons.org/licenses/by/4.0/).

DNA replication utilizing helicases for DNA unwinding (HDA). False positives arising out of these can be nullified by using specific oligonucleotide probes immobilized onto the electrode surface.

Electrochemical biosensors/immune-/genosensors are based on bioreceptors for recognition events and electrochemical transduction to transform the specific interactions into detectable electrical signals (Fig. 2). These sensing platforms provide a simple, rapid, economical, multiplexed, and portable format for the detection of cancer biomarkers (Campuzano et al., 2017a, d). Performance of electrochemical biosensors is largely determined by the amount, spacing, and stability of the bioreceptors immobilized on the surface of the sensing device, incorporation of a wide variety of unique nanomaterials into their fabrication with a large active surface area, abundant binding points, favorable catalytic activity, conductivity, and biocompatibility. This results in improving the sensitivity, selectivity, minimize interferences from complex biological matrices and prolong stability, thus meeting the demands of the detection of extremely low concentrated circulating biomarkers in scarcely treated samples (Pinto et al., 2013).

Nanomaterials owing to their unique features, such as noble metal and metal oxide nanoparticles (NPs), nanochains (NCs) and nanospheres, integrated nanostructures including graphene or reduced graphene oxide (rGO), or metallic NPs or oxides, multiwalled carbon nanotubes (MWCNTs), screen-printed electrodes (SPEs), magnetic beads (MBs), and biosensors incorporating above are capable of detecting clinical biomarkers with high sensitivity, and accuracy, and lower limit of detection (LODs). MBs facilitate isolation,

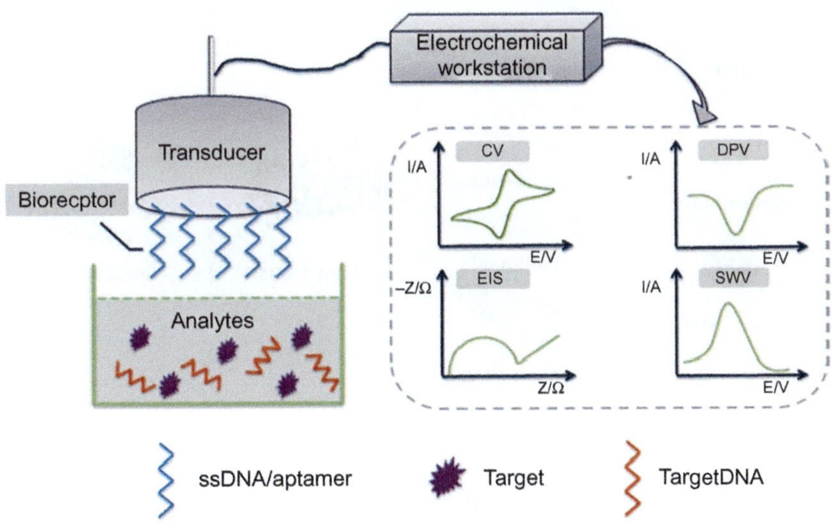

FIG. 2 Schematic diagram of electrochemical DNA biosensors (Wu et al., 2019). Copyright © 2019 by the authors. Licensee MDPI, Basel, Switzerland. This article is an open-access article distributed under the terms and conditions of the Creative Commons Attribution (CC BY) license (http://creativecommons.org/licenses/by/4.0/).

purification, and pre-concentration of the analytes of interest by offering a large surface area for immobilization, and solid support. This significantly improves the electrochemical biosensing performance in terms of sensitivity, reproducibility, and reduced assay turnaround. MBs are dispersible in solution, thereby enhancing the capture efficiency of cancer biomarkers. MBs are highly selective and specific as it enables the biorecognition and transduction steps on different surfaces thus reducing the chances of nonspecific adsorptions. SPEs are utilized for fabricating portable, low sample consuming electrochemical biosensors. The planar surface of SPEs aid in their modifications and linkage with MBs via a magnetic field (Pinto et al., 2013).

Nanomaterials such as graphene, rGO, AuNPs, AuNRs, MWCNTs incorporated nucleic acid impart large surface area with abundant binding points, catalytic activity, conductivity, and biocompatibility for electrochemical genosensing platforms with improved selectivity, sensitivity, signal-to-noise ratio, lowered detection limit, minimum interference from biological matrices and prolonged stability. Sensors based on electrochemiluminescent functional nucleic acids (ECL-FNAs) utilizes series of sensing devices that use FNAs as the recognition elements and ECL as the transducer. Nanomaterials can be used as carrier (graphene, metallic NPs), enhancer (metal NPs, carbon NMs, plasmonic NPs), catalyst (noble metal NPs, metallic oxide NPs, composite NPs), reporter (Metal NPs, metallic oxide NPs, QDs), quencher (AuNPs), separator (magnetic NPs of Fe, Ni, Co) and play other important roles in different nanobiosensing-based methods (Singh, 2014, 2016a, b; Xu et al., 2019) (Fig. 3).

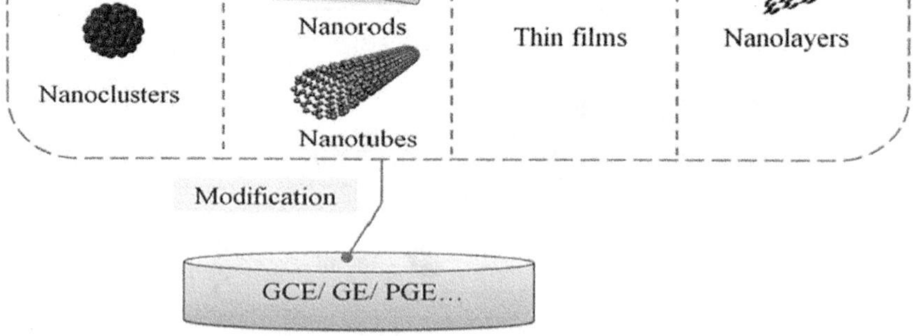

FIG. 3 Nanomaterials commonly used in the modification of electrochemical DNA biosensors (Wu et al., 2019).

3.2. Electrochemical immunosensing for the assessment of circulating biomarkers

This includes body fluids such as blood, urine, saliva, amniotic fluid, breast milk, bronchoalveolar lavage, cerebrospinal fluid (CSF), semen, pleural effusion, ascites, etc. (Campuzano et al., 2017c).

- **Risk screening biomarkers** are used to determine the disease susceptibility of an individual,
- **Prognostic biomarkers** provide information on the probability of the disease in an untreated individual,
- **Predictive biomarkers** are particular proteins or genes indicating sensitivity or resistance to a specific therapy, and
- **Diagnostic and disease monitoring biomarkers** used to aid the diagnosis and monitoring of the progression of disease (Campuzano et al., 2017c).

3.2.1 Clinical biomarkers

Determination of disease using several biomarkers on the same biosensing platform is more promising than the detection of a single disease-specific biomarker. Simultaneous determination of multiple biomarkers alleviates the chance of getting false positives and false negatives. Later is made possible by the innovation of multiplexed biosensing platforms having multiple transductors/detection labels for each analyte. The design of a multiplex sensor requires a distinct signal corresponding to each analyte from multiple ligand-analyte reactions. This could be achieved by using a multielectrode array where each electrode is assigned to each analyte. A barcode configuration can also be used where a unique electrode platform is provided for different electroactive labels with different electrochemical properties for each analyte emitting distinct signals with changing potential. These configurations use single labels thus avoiding cross-contamination from adjacent transducing elements. These sensors use solid carbon or metal electrodes incorporating graphene or CNTs, AuNPs, etc. for immunoreagent immobilization (Yáñez-Sedeño et al., 2019).

The need of the hour is the development of electrochemical sensing platforms that can detect genetic biomarkers (DNAs), regulatory biomarkers (RNAs), functional biomarkers (proteins), and metabolic biomarkers (small biomolecules) for simple, rapid, and reliable diagnosis of different pathological conditions, genetic, neurological, metabolic, and physiological disorders. Electrochemical techniques are advantageous for the detection of clinical biomarkers than the optical sensing techniques as the former offers reliability, precision, noninvasiveness, portability, the economy of cost, compatibility with POC devices, compatibility with microfabrication techniques, semiconductor technologies, and adaptability in a broad range of settings. The high sensitivity of electrochemical detection is due to the surface-bound probes and the pre-concentration step of analytes. However, the challenges remain in terms of validation in actual clinical samples, reproducibility, storage, transportation, and applicability in a resource-limited remote setup for the on-site detection (Campuzano et al., 2017a, d).

3.2.2 Multiplexed electrochemical immunosensors for the detection of cancer biomarkers

Electrochemical immunosensors are mostly sandwich-type having career labels attached to secondary antibodies. The electrode arrays are commonly fabricated with screen-printed electrodes (SPEs) (Costa Rama and Costa-García, 2016). Bode and Dong (2004) have reported the multiplexed sensing platform for the detection of phosphorylated p53 at Ser392 (phospho-$p53_{392}$), Ser15 (phospho-$p53_{15}$), Ser46 (phospho-$p53_{46}$), and total p53 as cancer biomarkers. This involved immobilizing different capture antibodies on each of the working electrodes and using Au-nanorods as nanocarriers for co-immobilization of horseradish peroxidase (HRP) and detection of antibodies. Another study reported disposable screen-printed carbon electrode (SPCE) array with 16-detection platforms and multi-labeled nanoprobe (HRP-anti-rabbit-IgG on MWCNT) for the multiplexed detection of prostate specific antigen (PSA), and interleukin 8 (IL-8) as cancer biomarkers using multiplex electrochemical immunosensor. A large number of HRP molecules on CNTs allow high sensitivity at lower detection limits (Yáñez-Sedeño et al., 2019). A multiplexed electrochemical immunosensor was designed for the simultaneous detection of phosphorylated proteins phospho-$p53_{15}$ and phospho-$p53_{392}$ using different apoferritin-templated SiO_2-AuNPs modified metal phosphates, i.e., cadmium (ATCP) and lead (ATLP). A multiplexed sandwich-type electrochemical immunosensor with PAMAM dendrimer-metal sulfide QDs as signal labels was designed for the simultaneous detection of carbohydrate antigen CA125, CA15-3, and CA19-9 cancer biomarkers employing stripping voltammetry (SV) (Yáñez-Sedeño et al., 2019) (Fig. 4).

Carbon nanospheres (CNSs) are produced by hydrothermal carbonization of carbohydrates and come up with features like abundant reactive oxygen functional groups, adsorption capacity, biocompatibility, and are utilized for the immobilization of signal probes and other biomolecules. A multiplexed electrochemical immunosensor was designed where, CNSs were coated with AgNPs or AuNPs and utilized glassy carbon electrode (GCE) modified with AuNPs/rGO nanocomposites for the simultaneous detection of carcinoembryonic antigen (CEA) and AFP (Cui et al., 2014; Ge et al., 2016; Li et al., 2016; Zhou et al., 2016; Yáñez-Sedeño et al., 2019). TiO_2NPs with features like biocompatibility, high reactivity, low cost, good conductivity, and electrocatalytic activity are used for constructing multiplexed electrochemical immunosensor in sandwich format for the determination of cancer biomarkers (CA15-5, CA19-9). This immunosensor utilized Nafion-TiO_2 nanohybrid for the immobilization of immunoreagents as label tags (Ensafi et al., 2014). Besides, polymer-NPs integrated with AuNPs were used for the construction of multiplexed electrochemical immunosensor for the detection of AFP and CEA biomarkers.

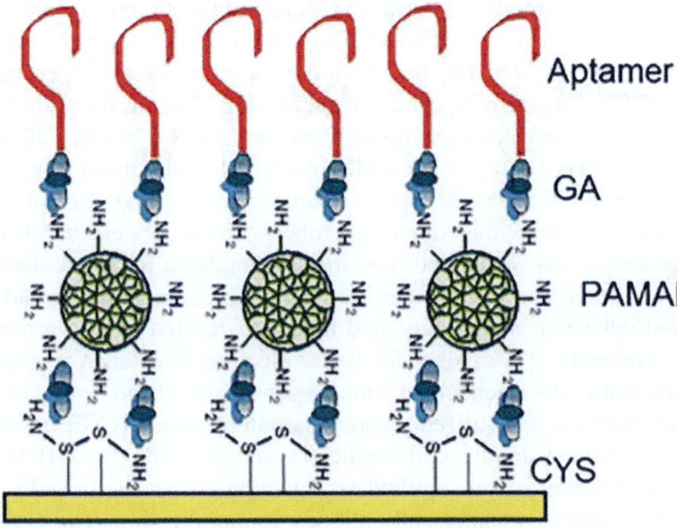

FIG. 4 The schemes of aptasensors: aptamers are immobilized at the surface of PAMAM G4 dendrimers (Karapetis et al., 2018). Copyright © 2018 by the authors. Licensee MDPI, Basel, Switzerland. This article is an open-access article distributed under the terms and conditions of the Creative Commons Attribution (CC BY) license (http://creativecommons.org/licenses/by/4.0/).

An immunosensing platform was developed for the detection of CA15-3 that does not require reagent and mediator. This platform utilized GCE modified with CNTs and nanospheres with bilayered Pt nanoclusters for glucose oxidase (GOD) and adsorbed silica. Results show that peak current corresponding to cofactor oxidation decreases with the increasing concentration of CA15-3 (Campuzano et al., 2017a, d). An ultrasensitive sandwich immunosensor was developed for the determination of interleukin 8 (IL-8). The immunosensing platform was having a sensor surface made up of glutathione-protected AuNP modified with capture antibody and super magnetic beads loaded with detector antibody. An electrochemical immunosensing platform was developed for the detection of carcinoembryonic antigen (CEA) utilizing a magnetic nanocomposite with redox activity and gold-graphene nano-labels. Gold-graphene nano-labels were used for facilitating AuNP immobilization and multiple biomolecules thus permitting antigen-antibody interaction. A label-free impedimetric immunosensor was developed for the detection of the epidermal growth factor receptor (EGFR). This utilized AuNPs modified with protein G and electrodeposited on an Au-electrode modified with a cysteamine self-assembled monolayer (SAM) (Campuzano et al., 2017a).

An electrochemical immunosensor was developed for the analysis of the HER2 extracellular domain (ECD) in human serum based on a sandwich assay having AuNP modified SPEs. Emami et al. (2014) have developed a label-free immunosensor for a similar purpose where thiolated antibodies were linked with PEG-maleimide-coated iron oxide nanoparticles (Fe_3O_4 NPs) resulting in bioconjugate formation to be deposited over Au-electrode. An electrochemical aptamer/antibody (Apt/Ab) sandwich immunosensor was developed for the detection of EGFR. This sensing platform successfully discriminated between healthy and breast cancer serum samples based on EGFR content. This sensing platform has utilized Strept-coated MBs modified with a biotinylated anti-human EGFR aptamer that acts as capture microcarriers while AuNPs linked to detector antibodies were used as signaling probes. A label-free electrochemical CEA immunosensor was developed that utilized graphene for good electrical conductivity and bioactive surface area, AuNPs for biocompatibility, and chitosan for good adhesion, nontoxicity, and abundant reactive amino groups to facilitate the synthesis of a cryogel for better electron transfer. A nanostructured, nonenzymatic electrochemical immunosensing platform was designed for CEA. This platform has utilized a glassy carbon electrode (GCE) modified with a nanocomposite of stannic oxide/reduced graphene oxide (SnO_2/rGO) and AuNPs for immobilizing capture antibodies. Nanocomposites facilitated a large surface area for electron transfer and antibody immobilization. The captured CEA was then used with PdNPs-vanadium pentoxide (V_2O_5)/MWCNTs and linked with secondary antibodies (Ab_2). The PdNPs-V_2O_5/MWCNTs were used for the catalytic reduction of H_2O_2 (Campuzano et al., 2017a, d).

A sandwich-type electrochemical CEA immunosensor was proposed utilizing an enzyme-free signal amplification strategy aided by graphene oxide/chitosan-ferrocene (GO/CS–Fc) as a matrix for immobilizing capture antibodies and Fe_3O_4/AuNPs functionalized detector antibodies. The CEA was determined by the catalytic reduction of p-nitrophenol to electroactive p-aminophenol using gold-nanocatalyst labels (Peng et al., 2016). A label-free CEA immunosensing platform was described that was based on Pd-Ir bimetallic NPs acting as a catalyst. This platform has utilized GCE modified with Pd-Ir bimetallic NPs and chitosan for capturing a specific antibody (Zheng et al., 2016). Another label-free CEA immunosensor was reported using AuNPs-decorated Prussian blue-poly (3,4-ethylenedioxythiophene) (AuNPs/PB-PEDOT) nanocomposite for biocompatibility and redox activity (Yang et al., 2017).

3.2.3 Mesoporous metallic structures

Mesoporous metallic structures possess interconnected hollow channels that allow mass transport and enhance conductivity. Mesoporous PtNPs (MPtNPs) acts as nonenzymatic labels for secondary antibodies in sandwich-type immunosensor array. Such platform was utilized for the detection of carbohydrate antigen 125 (CA125), CA153, and carcinoembryonic antigen (CEA) as tumor markers utilizing triple graphene-modified-SPCE (Cui et al., 2014). Multiplexed sensors with stripping analysis of AgNPs are available, however, their limited stability in saline buffers and tedious synthetic protocols are big deterrents. A multiplexed sandwich-type magneto-immunosensing platform utilizing dual

SPCEs was developed for the detection of estrogen and progesterone receptors in cell-lysates as cancer biomarkers.

Ordered mesoporous carbon (OMC) is characterized by the high specific surface area and the large pore volume. The incorporation of metals in ordered matrix results in (OMC-M), which is used as nanoprobes in an electrochemical immunoassay (Yáñez-Sedeño et al., 2019). For instance, Zn and Cd nanocrystallites can be incorporated into OMC to give rise to OMC-Zn and OMC-Cd. These were integrated with the secondary antibodies to be used as signal probes for designing a multiplexed sandwich immunosensor for the detection of AFP and human epidermal growth factor receptor type-2 (HER-2). This immunosensor has utilized an MWCNTs-Chit modified GCE for the immobilization of capture antibodies (Yáñez-Sedeño et al., 2019). Zhou et al. (2016) have constructed a simple dual-signal-labeled electrochemical immunosensor using rGO as the substrate deposited onto GCE to immobilize the antibodies. The signal probes used in this configuration were CdSeTe@CdS quantum dots (QDs) and Ag nanoclusters (AgNCs) incorporated into mesoporous silica nanospheres for the detection of B-cell lymphoma 2 (Bcl-2) and Bcl-2-associated X protein (Bax) as a biomarker for monitoring the apoptosis of tumors.

3.2.4 Redox mediators as career tags for the detection of cancer biomarkers

Prussian Blue (PB), toluidine blue (TB), ferrocene or ferrocene carboxylic acid (FCA), thionine, (THI) and anthraquinone (AQ), are used as carrier tags and redox mediators to prepare immunoconjugates for detection in various configurations. A multiplexed electrochemical immunosensor in the sandwich format utilizing carboxylated graphene nanosheets conjugated with TB or PB and chitosan modified GCE was designed for the detection of CEA and AFP. A multiplexed electrochemical immunosensor in a sandwich format having graphene-AuNPs hybrid deposited on GCE was designed for the simultaneous detection of CEA, AFP, CA125, and PSA (cancer biomarkers) (Yáñez-Sedeño et al., 2019). Yang et al. (2014) have designed a multiplexed immunosensor combining multi-label strategy and multi-spot assay with an array electrode (AuNP/GCE) for the simultaneous detection of six biomarkers for hepatocellular carcinoma (HCC).

3.2.5 Alternative nanomaterial-based strategies for the detection of cancer biomarkers

In a study, signal tags with distinct electrochemical behavior were prepared by complexing Cd^{2+} and Cu^{2+} with amino-capped platinum porous nanoparticles (PtPNPs). This was followed by conjugation with anti-CEA and anti-AFP antibodies respectively, used as distinguishable signal tags for capturing antigens. A GCE modified with an ionic liquid was used as the electrode substrate for immobilization of capture antibodies. The voltammetric signals monitored by differential pulse voltammetry (DPV) correspond to the concentration of the antigen. A carbon gold nanocomposite (CGN) with high adsorption capacity was prepared by carbonization of glucose and deposition of AuNP. This was used in sandwich type immunosensor for the detection of CEA, PSA, and AFP (Yáñez-Sedeño et al., 2019).

3.2.6 Microfluidic device assisted cancer biomarker detection

Microfluidics offers the advantages of small working volumes, short assay times, and high sensitivity. This has facilitated its application for the development of electrochemical immunoassay. A microfluidic-based electrochemical immunoassay was developed for the multiplexed detection of PSA and IL-6 in serum as a marker for prostate cancer. This involved the offline capture of analytes by HRP-labeled magnetic particles antibody bio-conjugates followed by the attachment of capture antibodies to a chip with 8-electrodes modified with glutathione decorated AuNPs (Chikkaveeraiah et al., 2011). A flow-through multiplexed immunoassay in sandwich format was designed for the simultaneous electrochemical detection of CEA and AFP. The detection platform has utilized magnetic graphene-nanosheets with immobilized capture antibodies as immunosensing probes, and multifunctional nanogold hollow microspheres (GHS) with encapsulated HRP-THI and HRP-Fc as distinguishable signal tags. An electrochemical immunosensor utilizing microfluidic paper-based analytical devices and μ-PADs technology was developed for the simultaneous detection of biomarkers. This platform utilized nanomaterials for enhancing the sensor probe immobilization and sensitivity of detection. A 3D microfluidic electrochemical origami immunosensing platform was developed for the simultaneous detection of CA125 and CA19-9 biomarkers. Sun et al. (2015) have constructed a microfluidic paper-based sandwich immunoassay using μ-PAD in which rGO was used to modify the cellulose fibers to determine HCG, PSA, and CEA. ZnO-nanorods with a large surface-to-volume ratio and abundant sites for capturing biomolecules were employed to immobilize the specific antibodies.

3.2.7 Nucleic acid-based electrochemical genosensing of circulating biomarkers

Electrochemical genosensors are nucleic-acid-based electrochemical biosensors. It employs sequence-specific probes as a biorecognition element and electrochemical transduction to convert nucleic acid hybridization into a detectable signal. Electrochemical genosensors have demonstrated great potential to circumvent the deficiencies of conventional technologies for the determination of circulating biomarkers associated with various clinical morbidities (Castañeda et al., 2007) (Fig. 5).

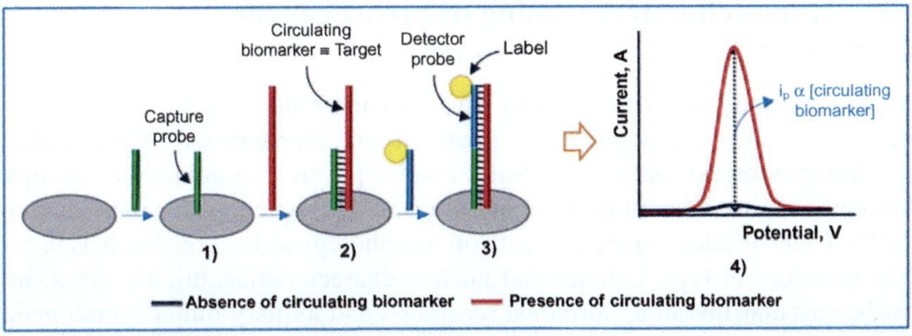

FIG. 5 Schematic display of a general nucleic acid-based genosensor: (1) specific hybridization of the immobilized capture probe with the target sequence (2) and of the captured target with the label-conjugated detector probe (3) and electrochemical detection of the hybridization reactions by differential pulse voltammetry of the label attached to the detector probe (4) (Campuzano et al., 2017a).

High-affinity bioreceptors such as ss-DNA, RNA sequences, aptamers, peptides, magnetic beads (MBs), screen-printed electrodes (SPEs), and nanomaterials are crucial in improving the analytical performance of electrochemical genosensors in terms of portability and cost-reduction. MBs considerably enhance electrochemical genosensing in terms of increased sensitivity, reproducibility, and reduced assay turn-around (Low et al., 2015). MB-based genosensing protocols perform hybridization and transduction steps on different surfaces thus minimizing nonspecific adsorption, increasing the sensitivity and specificity. SPEs with planar structure allows easy surface modifications in addition to portability, low sample consumption, the versatility of design, customization, and multiplexed analysis (Costa Rama and Costa-García, 2016).

Chen et al., 2011 have designed an electrochemical amperometric genosensor based on nuclease-assisted target recycling and DNAzyme for sensitive and selective determination of oral cancer overexpressed 1 (ORAOV1)-related DNA species. Similar biomarkers were detected by immobilization-free electrochemical genosensor. In another study, direct detection of epidermal growth factor receptor (EGFR) gene in saliva samples via multiplexed electrochemical genosensor in sandwich format was developed. Another sandwich-type electrochemical chronoamperometric genosensor was employed for the detection of BRAC1, a cancer biomarker utilizing rGO/GCE. Pingarrón's group 2015 has designed an electrochemical DNA biosensor utilizing screen printed carbon electrodes (SPCEs) modified with a reduced graphene oxide-carboxymethylcellulose hybrid nanomaterial for the detection of a target DNA sequence of the *p53* tumor suppressor gene.

3.3. Electrochemical sensing of breast cancer biomarkers

Breast cancer is one of the leading causes of mortality in females. Early and precise diagnosis is crucial for the treatment and management of the disease. Conventional breast cancer diagnostics involve physical examinations, histopathology, imaging mammography, and ultrasound magnetic resonance imaging (MRI). Conventional approaches rely on morphological features, such as tumor size, histological type, cellular and nuclear characteristics, mitotic index, necrosis, vascular invasion, hormonal receptors, and axillary tumor lymph node. However, these do not provide very precise screening details. Molecular biology led the discovery of novel biomarkers like DNA, mRNA, miRNA, exosomes, cell surface receptors, transcription factors, and secreted proteins in liquid biopsies are extremely valuable for the minimally invasive cancer diagnosis (Campuzano et al., 2017a, d). However, these techniques are fraught with limitations such as the inability of ultrasound to detect a variety of cancers, limited sensitivity and false positives of mammography, and invasiveness of histopathology. These limitations warrant urgent address of these issues and development of simple, rapid, economical, portable, noninvasive, and ultrasensitive POC screening of breast cancer. Electrochemical biosensing approaches by and large suit the need for cancer screening/diagnosis.

The advent of nanomaterials equipped with unique features, such as noble metal and metal oxide nanoparticles (NPs), nanochains (NCs), nanospheres, integrated nanostructures including graphene or reduced graphene oxide (rGO), metallic NPs or oxides, multiwalled carbon nanotubes (MWCNTs), screen-printed electrodes (SPEs), magnetic beads (MBs) play important role in the development of biosensors capable of detecting biomarkers for breast cancer with high sensitivity, accuracy with a lower limit of detection (LODs). MBs provide a large surface area for immobilization and solid support for the isolation, purification, and pre-concentration of the analytes of interest thus significantly improving the electrochemical biosensing performance in terms of sensitivity, reproducibility, and reduced assay turnaround. Being dispersible in solution MBs enhance the capture efficiency of cancer biomarkers. MBs enhance the selectivity and specificity by carrying out the biorecognition and transduction steps on different surfaces thus reducing the chances of nonspecific adsorptions. SPEs are used for portable, low sample consuming electrochemical biosensors. The planar surface of SPEs aid in the modifications and linkage with MBs via a magnetic field (Pinto et al., 2013).

3.3.1 Electrochemical biosensing of gene-specific mutations and miRNAs associated with breast cancer in biofluids

Genetic biomarkers such as mutations in specific genes (BRCA1, BRCA2, and p53 genes) and dysregulation measurement of specific miRNAs are noninvasive (Campuzano et al., 2017a, d). Benvidi et al. (2015) have devised a simple, reproducible impedimetric genosensor for the analysis of genomic DNA isolated from peripheral blood samples. The method involved the immobilization of a binary self-assembled monolayer (SAM) composed of a thiolated specific capture DNA probe (BRCA1 5382 insC mutation detection) and 6-mercapto-1-hexanol (MCH) on a gold electrode. Wang et al. (2017) have developed an impedimetric DNA hybridization biosensor for detecting BRCA1 mutations. Another sandwich-type electrochemical chronoamperometric genosensor was employed for the detection of BRAC1, a cancer biomarker utilizing rGO/GCE nanocomposite (Campuzano et al., 2017a).

miRNA biomarkers are noninvasive and are largely responsible for the cancer initiation, progression, and tumor response to the treatments. These are stable in serum/plasma and their levels are specifically dysregulated in association with cancer. Amplification of chromosomal regions encoding oncogenic miRNAs (miR-21 and miR-155, inhibiting tumor suppressor genes) leads to the upregulation of miRNAs and silencing of cancer developing genes. Tumor suppressor miRNA (miR-205) is located in the fragile regions of chromosomes. The mutation or chromosomal rearrangement of these regions results in the loss of expression of these miRNAs and up-regulation of their target genes (Campuzano et al., 2017a, d). A biosensing platform was developed for the detection of miRNA-21. The target miRNA was sandwiched between a capture probe (CP) immobilized onto the gold electrode surface and a methylene blue (MB, redox mediator) labeled G-rich detection probe (DP) modified onto AuNPs. The study has employed G-quadruplex DNA species and iridium(III) complex as peroxidase to act as a mimic to catalyze the reduction of H_2O_2. Shuai et al. (2016) have developed an ultrasensitive electrochemical biosensor for miRNA-21. This was based on tungsten oxide-graphene composites (WO_3-Gr) coupled with catalyzed hairpin assembly target recycling and enzyme signal amplification. Electrochemical biosensing platforms for detecting protein biomarkers for cancer utilize antibodies and aptamers or both.

An integrated electrochemical sensing platform was developed for the simultaneous detection of IL-8 mRNA and IL-8 protein in the same assay. This involved immobilizing the biotinylated specific bioreceptors—hairpin probe sequence and antibody onto a 16 gold electrodes-array chip. This was modified with a conducting polymer incorporating Strept-dendrimer nanoparticles to improve the biocompatibility allowing for chronoamperometric measurements. In another study magnetic bead based dual amperometric platform was developed for IL-8 mRNA and IL-8 protein detection. This platform utilized functionalized Strept-MBs and HOOC-functionalized magnetic beads specific antibodies against IL-8 protein, a specific biotinylated hairpin DNA sequence for IL-8 mRNA, and amperometric detection at disposable screen-printed dual carbon electrodes (SPdCEs) (Campuzano et al., 2017a, d).

3.3.2 Electrochemical Aptasensors for breast cancer protein circulating biomarkers

Zhao et al. (2011) have devised a folding-based electrochemical DNA aptasensor for the detection of vascular endothelial growth factor (VEGF) directly in blood serum and whole blood. The protocol involved the immobilization of the thiol and methylene group labeled aptamer on a gold electrode. The electrochemical signal of MB, measured by alternating current voltammetry (ACV) was enhanced in the presence of VEGF indicating the formation of a rigid stem-loop structure in the aptamer bringing the MB label closer to the electrode. Salimian et al. (2017) have devised a label-free aptameric sensor for human epidermal growth factor receptor 2 (HER2). This aptasensor utilized the co-adsorption of a specific thiolated aptamer sequence on gold along with a C11 alkane-thiol bearing two ethylene glycol $(EG)_2$ head groups to prevent nonspecific adsorptions of nontarget proteins. The measurement of the electrocatalytic reduction of ferricyanide redox indicator by MB electrostatically interacting with negatively charged HER2 allowed the detection of serum.

A sandwich-type amperometric magneto-immunosensors was developed by Eletxigerra et al. (2016) for the determination of HER2 and estrogen receptor α (ERα) in human serum with the ability to discriminate breast cancer cells with respect to the expression level of the target receptor. In both cases, capture antibodies were covalently immobilized on carboxylic acid-modified MBs and the captured target protein sandwiched with HRP-labeled (HER2) or biotin-labeled (ERα) detector antibodies further conjugated with a Strept-HRP polymer. The sensing platform was captured on screen-printed carbon electrodes (SPCE) and the cathodic current measured by amperometry was directly proportional to the concentration of target proteins.

3.3.3 Electrochemical peptide-biosensor for the detection of circulating breast cancer protein biomarkers

An impedimetric biosensor was reported for the determination of VEGF utilizing vascular endothelial growth factor receptor-1 (VEGF-R1) as the recognition element covalently immobilized onto a gold electrode modified with an MPA SAM. An electrochemical biosensing platform was developed for the determination of HER2 protein and HER2-overexpressing breast cancer cells. The sensor was composed of a covalently immobilized specific capture antibodies onto a nanocomposite layer composed of self-assembled 2,5-bis(2-thienyl)-1H-pyrrole-1-(p-benzoic acid) (DPB) on AuNPs and Hydrazine-AuNPs-aptamer bioconjugates as labels for signal amplification. MAb-aptamer combined was responsible for the high sensitivity and selectivity. An electrochemical cytosensor based on a folate-linked DNA probe was reported for the detection of human breast cancer MCF-7 cells. This protocol was based on the digestion of folate-linked DNA probes immobilized on the electrode surface by exonuclease I (Campuzano et al., 2017a, d). Wang et al. (2017) have developed a sandwich electrochemical aptasensor for the label-free and selective detection of breast cancer cells (MCF-7). This aptasensor utilized a polyadenine (polydA)-aptamer modified gold electrode and a polydA-aptamer functionalized AuNPs/GO hybrid. The decrease in the $[Fe(CN)_6]^{3-/4-}$ DPV peak current in the presence of the sandwich system was employed to quantify the breast cancer cells. The high selectivity was ensured by the use of aptamer, which works against the MUC1 protein, overexpressed in MCF-7 cells.

3.3.4 Electrochemical biosensing for multiple determination of circulating breast cancer biomarkers

Multiplexed detection of specific cancer biomarkers is a reliable technique free of false positives for the diagnosis and prognosis of breast cancer as against the single biomarker. This is due to the population variations in the expression of a single biomarker, lack of specific biomarkers, and the multiple complex molecular events involved in breast cancer development (Lin et al., 2016). A **dual scaffold sensing platform** was developed by coupling two previously developed single strategies for the simultaneous detection of progesterone receptor (PR) and ERα in cellular lysates and spiked human serum samples at SPdCEs. The biosensor design utilized two different batches of functionalized magnetic beads bearing HRP-labeled sandwich immunocomplexes specific for each protein receptor, which were magnetically captured on the corresponding working electrode of the SPdCE (Campuzano et al., 2017a, d).

3.4. Electrochemical biosensor for the detection of prostate cancer biomarkers

A microfabricated glass chip-based electrochemical biosensor was developed for counting prostate cancer cells (DU145 cells) in a complex media with serum or a mixed cell population containing nontarget cells. This platform facilitated simple, rapid, label-free detection of prostate cancer cells with high selectivity and sensitivity. The sensing platform was equipped with a gold electrode array with tunable sensor surface areas. The sensor surface was modified with a capture antibody specific for epithelial cell adhesion molecule (EpCAM, anti-EpCAM antibody) to selectively count prostate cancer cells using the differential pulse voltammetric (DPV) response in the presence of $[Fe(CN)_6]^{3-/4-}$ (Campuzano et al., 2017a). Another study reported a disposable screen-printed carbon electrode (SPCE) array with the 16-detection platform and multi-labeled nanoprobe (HRP-anti-rabbit-IgG on MWCNT) for the detection of multiplexed electrochemical immunosensor for the detection of prostate-specific antigen (PSA) and interleukin 8 (IL-8) as cancer biomarkers.

A microfluidic-based electrochemical immunoassay was developed for the multiplexed detection of PSA and IL-6 in serum as a marker for prostate cancer. This involved the offline capture of analytes by HRP-labeled magnetic particles antibody bio-conjugates followed by the attachment of the capture antibodies to a chip with 8-electrodes modified with glutathione decorated AuNPs (Chikkaveeraiah et al., 2011).

3.5. Biomarkers for cardiovascular disease

Nanoscale electrochemical biosensors are suitable for the early detection of cardiovascular disease (CVD) biomarkers at very low concentrations in biological fluids. Different CVD biomarkers include AXL receptor tyrosine kinase, apolipoproteins, cholesterol, C-reactive protein (CRP), D-dimer, fibrinogen (Fib), glucose, insulin, interleukins, lipoproteins, myoglobin, N-terminal pro-B-type natriuretic peptide (BNP), tumor necrosis factor-alpha (TNF-α), and troponins (Tns) on electrochemical transduction format. These offer high sensitivity, affinity, and selectivity for bioreceptor and stability owing to nanostructuring the electrode surface (Farzin et al., 2018). Cardiac markers such as cardiac troponins (cTnT, cTnI) or myoglobin (Mb) serves as an index of acute myocardial infarction (AMI), while cardiac reactive protein (CRP), B-type natriuretic peptide (BNP), and N-terminal ProBNP act as an indicator for systemic inflammations.

3.5.1 Electrochemical immunosensing of cardiovascular disease biomarkers

A multiplexed sandwich-type magnetic-microparticle-based immunosensor was designed for the simultaneous detection of NT-proBNP and CRP. The results were in sync with the clinical findings (Esteban-Fernández de Ávila et al., 2015). Gupta et al. (2016) designed an immunosensing array configuration with a multielectrode sensor chip having nine identical, electrically isolated, microelectrodes for the immobilization of immunoreagents. This platform was used for the simultaneous detection of CRP, cTnI, and Mb cardiac biomarkers. Another integrated chip (silicon nanowire) based immunosensor was designed that utilized finger-prick blood for the detection of cardiac biomarkers like cTnT, creatine kinase MM (CK-MM), and creatine kinase MB (CK-MB) (Zhang et al., 2011). A multiplexed immunosensing platform was developed utilizing dual SPCEs modified with double-walled CNTs for the simultaneous detection of TNF-α and IL-1β cytokine biomarkers (Sánchez-Tirado et al., 2017). Similarly, a genosensor was designed for the detection of miRNA as a biomarker for cancer and cardiovascular disease. The protocol involved direct or competitive hybridization assays using RNA-RNA, DNA-DNA duplexes as specific recognition sequences.

3.6. Electrochemical biosensing of other disease biomarkers

3.6.1 Diabetes

A multiplexed immunosensor using dual SPEC modified with reduced graphene oxide (rGO) was used for the detection of PYY and GHRL in biological fluids as markers for diabetes (Verma et al., 2011).

3.6.2 Genetic disorder

A dual multiplexed electrochemical immunosensor having aminophenylboronic acid-modified dual SPCEs was developed for the detection of ACTH and cortisol as the marker for Cushing's syndrome (Moreno-Guzmán et al., 2012).

3.7. Electrochemical genosensors for the neurodegenerative disease biomarkers

Miodek et al. (2013) have constructed an electrochemical aptasensor for the detection of human cellular prions (PrPC) responsible for the transmissible spongiform encephalopathies (TSEs), a group of fatal neurodegenerative disease including Creutzfeldt-Jakob disease in human and spongiform encephalopathy (SE) in animals. This involved immobilizing specific biotinylated aptamer through streptavidin binding to a biotinylated scaffold (Collinge, 2011).

In another approach, a novel ratiometric, on-off signal mode electrochemical genosensor was designed for the detection of the mutated *apolipoprotein E* gene associated with Alzheimer's disease. The protocol utilized graphene and mesoporous hybrid nanomaterials (GSHs)-modified GCE. The GSHs served as nano-reservoirs for the electro-active molecules like MB. The binding of nontarget dsDNA to GSH surface in the absence of target DNA has led to the closing of nanopores while the presence of target DNA led to the opening of nanopores causing the MB molecules to release a signal output (Wu et al., 2016).

3.8. Electrochemical sensors for the detection and biomonitoring of viral and bacterial pathogenic biomarkers

Pathogenic infections especially, bacterial and viral infections are the major cause of mortality in humans globally. These infections are further aggravated by the wrong and delayed diagnosis that leads to delayed or wrong treatment. Simple, reliable, and rapid diagnosis of the causative clinical biomarkers of the infection plays an important role in controlling and managing antibiotic resistance and optimizing antibiotic administration. A timely and accurate on-site diagnosis utilizing simple, rapid, and economical protocols, easy translation to point-of-care-testing (POCT) is crucial for the effective treatment, control, and management of viral and bacterial diseases. Minimally invasive or noninvasive diagnosis may lead to effective therapy, and simultaneously reduces the quantum of procedure-associated complications (Ahmed et al., 2014). Conventional tests such as cell-culture based or serological tests are slow and time-intensive taking few days to weeks while nucleic acid-based molecular methods such as RT-PCR, ELISA, immunological techniques, etc. are rapid but the cost of infrastructural set-up coupled with skilled and trained manpower requirement, and low sensitivity are big deterrents in resource-limited settings of remote areas. Electrochemical molecular biosensors meet most of these requirements as these facilitate simple on-site detection, which is rapid and cost-effective. Electrochemical biosensors offer inexpensive, field-portable, and programmable battery-operated instrumentation as an exciting alternative for rapid and economical analytical strategies that can be easily handled by unskilled personnel. Affinity based electrochemical biosensors/immunosensors facilitates simple, rapid, economical, and minimally invasive protocol for the detection of circulating biomarkers in biofluids such as blood, serum, saliva, urine, sputum, and pleural fluid (for references, Campuzano et al., 2017d).

Electrochemical biosensors enable simple, rapid, cost-effective, low sample requiring, multiplexed, portable, minimal pre-treatment requiring detection strategies, which makes them the favorable choice for POC applications (Pan et al., 2010). Nanomaterials such as AuNP, CNT owing to their unique features like large surface area, multiple binding sites, conductivity and biocompatibility, sensitivity, selectivity, prolonged stability, and minimum interference from biological matrices are preferred for the fabrication of affinity-based electrochemical biosensors. These sensors utilize ss-oligonucleotide, peptide sequences, and antibodies. These sensing platforms are conjugated with different nucleic acid amplification protocols, such as PCR, loop-mediated amplification (LAMP), rolling-circle amplification (RCA), and mimicking in vivo DNA replication utilizing helicases for DNA unwinding (HDA) to achieve the required sensitivity. Specific oligonucleotide probes immobilized onto the electrode surface eliminates the false positives arising out of these protocols.

3.8.1 Viral disease biomarkers

In the case of viral pathogens, the infectious dose or the viral load is the number (number of viral particles per microliter of body fluid) of viruses present in the body fluids. Different diagnostic methods are considered for designing effective and reliable diagnostics platforms. Viral infections are diagnosed by determining the specific nucleic acid sequence of the target virus. Viral RNAs are usually detected between 10^3 and 10^6 copies/ml in biological samples. For Zika virus this range is 10^2 to 10^5 copies/ml in biological samples, while for Flavivirus influenza this range is 10^6 to 10^9 in an aspirate sample and 10^3 in the throat and nasal swabs. However, these levels are few thousand higher in sexual fluid like semen, etc. thus indicating the capacity for the sexual transmission of the infection (for references, Campuzano et al., 2017d; Singh, 2017a, b).

Virus-specific electrochemical immunosensors usually detect the whole virus (human enterovirus), specific viral proteins (p24, NS1 from HIV and Dengue respectively), auto-antibodies against specific viral antigens (pseudorabies virus, HIV-1, and HIV-2, hemagglutinin HA from avian influenza virus H5N1), and specific DNA sequences (Dengue virus serotype 3, DENV-3, high-risk human papillomavirus hrHPV, strains HPV-16 and HPV-18) (Campuzano et al., 2017d).

3.8.2 Electrochemical genosensing of biomarkers for viral infections

A protocol involving the use of specific DNA and peptide sequences has been reported recently for electrochemical genosensing of viruses and autoantibodies specific to them. An electrochemical peptide-based sensor was developed for the detection of HIV antibodies involving self-assembly of MB-labeled thiolated peptide as an indicator and short thiolated DNA as anti-fouling diluents on Au-electrode (Fig. 6). Oliveira et al. (2015) have constructed an electrochemical genosensor for the detection of dengue virus serotype 3 (DENV-3) utilizing a specific 22-mer DNA probe adsorbed onto a pencil graphite electrode using differential pulse voltammetry (DPV). A multiplexed electrochemical immunosensor based on the microfluidic paper array with 8-SPEs was constructed for the simultaneous detection of HIV and HCV viruses in serum employing

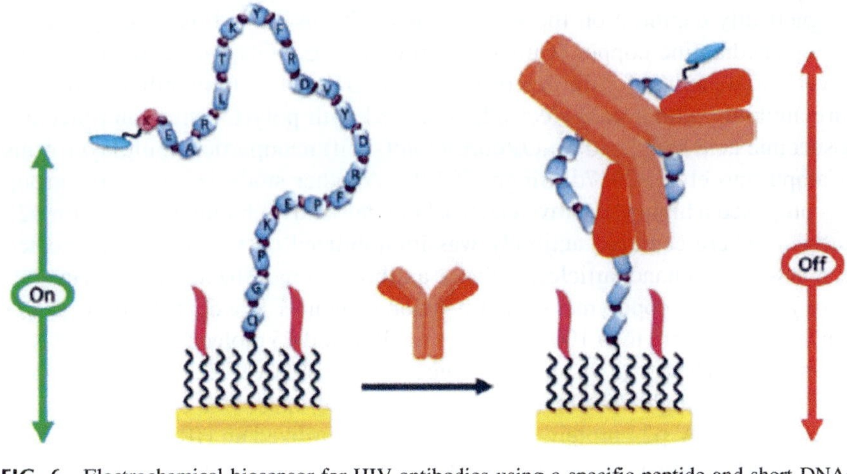

FIG. 6 Electrochemical biosensor for HIV antibodies using a specific peptide and short DNAs as diluents (Campuzano et al., 2017c).

amperometric detection of an electrochemical signal (Zhao and Liu, 2016; Singh, 2017a). In another study, human norovirus was detected using an impedimetric biosensor by assembling a thiolated affinity peptide on a gold electrode (Hwang et al., 2017).

Chandra et al. (2019) have studied the interaction between HAdV-D37 fiber knob: GAG using SPR. The group has demonstrated that fiber knob interacts with a broad range of sulfated GAGs. They had also shown that sulfated glycosaminoglycans (GAGs) can be used as decoy receptors for species D of human adenovirus type 37 (HAdV-D37), which is a major cause of epidemic kerato-conjunctivitis (EKC). This study has further established that the binding of HAdV-D37 with sulfated GAGs in secretions or on plasma membrane prevents or delays the viral binding to sialic acid-containing receptors and the subsequent infection. The study further established that the abundant heparan sulfate in the basement membrane of human corneal epithelium acts as a barrier to sub-epithelial infection (Singh, 2017b). Human papillomavirus (HPV) was detected using an electrochemical genosensor in hybridization format by immobilizing a specific capture probe on a GCE modified with graphene/Au nanorods/polythionine.

3.8.3 Minimally invasive electrochemical immunosensing of human immunodeficiency virus (HIV)

A label-free amperometric immunosensor was developed for the rapid determination of HIV p24 protein from the serum of HIV patients. This study has used magnetic bioconjugates of a specific antibody and Fe_3O_4 (core)/Au (shell) nanoparticles-coated multiwalled carbon nanotubes (MWCNTs). These were magnetically captured on the surface of *N, N*JJ-bis-(2-hydroxy-methylene)-*o*-phenylenediamine copper (CuRb)-modified SPCEs. Gan and coworkers have developed a method for the determination of HIV p24 by immobilizing the capture antibody onto a gold electrode modified with poly(L-lysine) and mercaptosuccinic acid stabilized Fe_3O_4(core)/gold(shell) nanoparticle multilayer films (Campuzano et al., 2017d; Singh, 2017b). Another study led by Gan's group has proposed a highly sensitive electrochemical sandwich immunoassay for p24 antigen, where captured antibody was immobilized on silicon dioxide-coated magnetic Fe_3O_4 nanoparticles; detector antibody immobilized on gold nanocolloids and a novel copolymer of an EnVision reagent (EV, a dextrin amine skeleton anchoring more than 100 molecules of HRP and 15 molecules of antiIgG) as a signal tag. The sensitivity was enhanced 1000-fold over ELISA (Campuzano et al., 2017d; Singh, 2017b). A sandwich immunosensor was developed for the detection of HIV p24 protein. This platform has utilized an HRP-labeled detector antibody and electroplating of AuNPs onto a GCE surface by chronoamperometry to enhance the conductivity of the electrode. This has facilitated the immobilization of a large number of capture antibodies while keeping their biological activity intact. In another study, a simple electrochemical immunoassay was developed for the detection of HIV-1 and HIV-2 antibodies where HIV-1 and HIV-2 antigenic peptides were covalently attached with a SU-8 substrate (a negative epoxy-based photoresist functionalized with biomolecules without any pretreatment due to the presence of exposed epoxy groups). This was followed by capturing AP-conjugated secondary antibody labeled-HIV antibodies and the oxidation of p-aminophenol generated by hydrolysis of p-aminophenyl phosphate by AP measured by differential pulse voltammetry (DPV). A peptide-based electrochemical immunosensor was developed for the detection of HIV antibodies in spiked saliva samples. This study has utilized a thiolated peptide labeled with methylene blue (MB) and short thiolated DNAs on gold electrodes (for references, Campuzano et al., 2017d; Singh, 2017b).

3.8.4 Pseudorabies virus (PRV)

An MBs-based electrochemical immunosensor was developed for the detection of pseudorabies virus (PRV) antibodies. The study has involved PRV-modified MBs used for the selective capture of secondary antibody labeled specific antibodies in swine serum (for references, Campuzano et al., 2017d; Singh, 2017b).

3.8.5 Influenza (flu) virus

An integrated immunosensor was developed for the detection of antibodies against hemagglutinin (HA) from avian influenza virus H5N1 with an ability to differentiate between vaccinated and nonvaccinated chicken sera. The protocol exploits immunorecognition of the recombinant His-tagged HA (His6-H5 HA) by anti-His IgG monoclonal antibody immobilized onto ProtA-modified GCE. The integrated immunosensor displays a 10^4 fold improvement in sensitivity over ELISA (for references, Campuzano et al., 2017d; Singh, 2017b).

3.8.6 Dengue virus

A study led by the Darwish et al. (2016) has developed a sensitive, label-free electrochemical immunoassay for the determination of dengue virus biomarker (NS1). This involved an immunosensing scaffold as indium tin oxide (ITO) electrode modified with aryl-diazonium-cation-derived zwitterionic antifouling molecules (4-sulfophenyl, 4-trimethylammoniophenyl, and 1,4-phenylenediamine) capture antibodies and AuNPs. This immunosensor exploits the rate of electron transfer resistance after antigen binding by electrochemical impedance spectroscopy (EIS) in the presence of $Fe(CN)_6^{3-}/Fe(CN)_6^{4-}$ for the measurement. A DNA-based label-free electrochemical biosensor was developed for the detection of specific dengue virus serotype 3 (DENV-3). Sequences were developed by adsorbing a 22-nucleotide DNA probe on a pencil graphite electrode (PGE) (for references, Campuzano et al., 2017d; Singh, 2017b).

3.8.7 Human enterovirus 71 (EV71)

An electrochemical immunosensor was developed for the detection of human enterovirus 71 (EV71), the causative agent of hand, foot, and mouth disease (HFMD) among children. The EV71 virions from the sample were selectively captured on AuNP coated ITO electrode modified with a specific monoclonal antibody labeled with magnetic nanobeads modified with the detector antibody and HRP (Campuzano et al., 2017d; Singh, 2017b).

3.8.8 Human papillomavirus (hrHPV)

An MBs-based sandwich hybridization approach was developed for the detection of DNA sequences from high-risk human papillomavirus (hrHPV) strains (hrHPV-16 and hrHPV-18). Here, target hrHPV-DNA is captured using the specific DNA capture probe-modified MBs sandwiched with a digoxigenin-labeled detector probe. The resulting sandwich duplexes were further labeled with an antidigoxigenin-HRP antibody. Upon magnetic capture of the resulting modified MBs, chronoamperometric detection was performed. The protocol facilitated selective differentiation between hrHPV-16 and hrHPV-18, analysis of cancer cell lines, and cervical brush smears of patients (for references, Campuzano et al., 2017d; Singh, 2017b).

3.8.9 Human norovirus

A highly sensitive impedimetric biosensor was developed for the detection of human norovirus. This involved assembling a thiolated affinity peptide on a gold electrode. A similar protocol was applied for the analysis of fetal bovine serum as well (Campuzano et al., 2017d; Singh, 2017b).

3.9. Bacterial biomarkers

The quantum of bacteria present in body fluids constitutes an infectious dose. Various diagnostic strategies take these into account for designing effective and reliable diagnostics. Bacteria including drug-resistant forms are detected by the presence of colonies. Typically, 1–10 colony-forming units (cpu)/ml of blood are enough to cause infection. For rRNA, the copy number of 10^3 or 10^4 corresponds to a sub-femtomolar concentration of RNA along with an excess of nontarget RNA. TB bacilli are present in 10^7–10^8 cells/ml of sputum. However, its viscosity makes it difficult to perform the diagnostic procedure. The DNA is extracted from the sputum for molecular analysis. This limits the direct detection as molecular-level analysis has to rely on bacterial culture (for references, Campuzano et al., 2017d; Singh, 2017a, b).

3.9.1 Electrochemical genosensors for bacterial infection biomarkers

A chronoamperometric genosensor in hybridization format was reported for the determination of *Escherichia coli* using an electrode array. The finding was further validated with uropathogenic clinical isolates. *Streptococcus pneumonia* was detected on a disposable amperometric MB-based genosensor coupled to asymmetric PCR (aPCR) (Fig. 7). This protocol enabled successful discrimination against closely related species of *Streptococcus*, i.e., *S. mittis* and *S. pneumonia* (Campuzano et al., 2017c). Detection of the causative organism of periodontal disease, *Porphyromonas gingivalis* was performed via electrochemical genosensor utilizing an electrochemical intercalator mixed with amplicons from direct PCR (dPCR). Signal was obtained in the form of indicator peak current measured by linear sweep voltammetry (LSV). *Schistosoma haematobium,* a causative organism of urogenital schistosomiasis was detected via an electrochemical genosensor in a sandwich format. The protocol involved sandwich hybridization of 16S rRNA via thiolated DNA capture probe assembled on Au-electrode and a detector probe labeled with fluorescein (FITC) (Campuzano et al., 2017c). An electrochemical genosensor was developed for the detection of the *rpoB* gene, a marker for drug-resistant tuberculosis using amplicons from clinical samples. This protocol successfully discriminated single nucleotide polymorphism (SNPs) responsible for drug resistance. An electrochemical genomagnetic sensor was developed for the chronoamperometric detection of insertion sequence *IS6110*, a marker for *M. tuberculosis*. A microfluidic-multiplexed

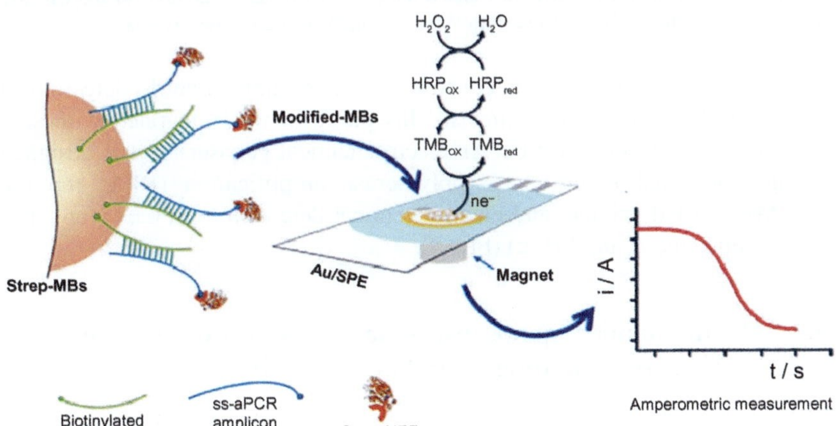

FIG. 7 Schematic display of the MBs-based amperometric DNA genosensor developed for *S. pneumoniae* determination (Campuzano et al., 2017c).

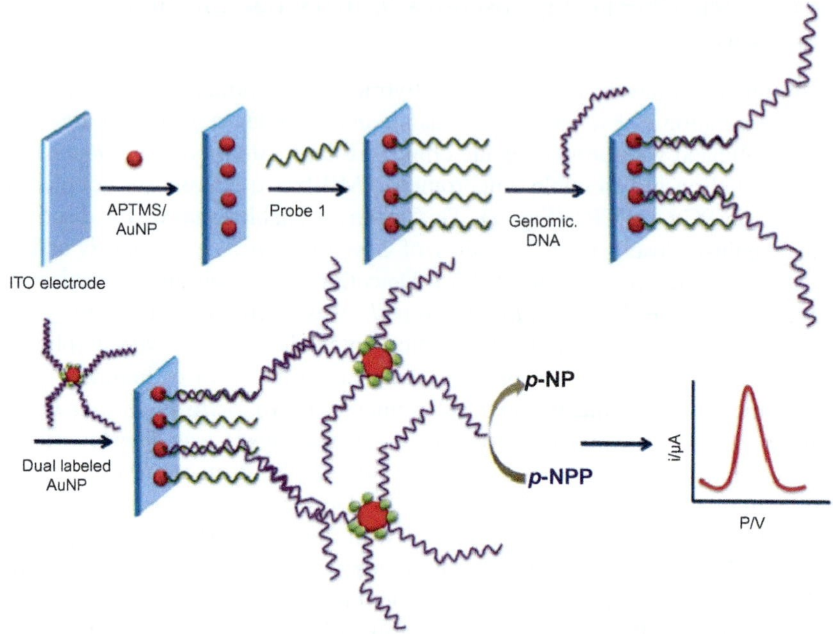

FIG. 8 Schematic display of the electrochemical DNA genobiosensor developed for *Mycobacterium* sp. genomic DNA detection using a specific probe immobilized on an ITO electrode and AuNPs modified both with detector probes and the enzyme alkaline phosphatase (AP) (Campuzano et al., 2017c).

electrochemical genosensor was constructed for the impedimetric detection of genomic DNA from *M. tuberculosis*. This platform was integrated with sensors composed of CNTs with Fc. An electrochemical genosensor was designed utilizing isothermal recombinase polymerase amplification (RPA) through AuNP-MB labeled primers and electrochemical detection at SPCEs (for references, Campuzano et al., 2017c) (Fig. 8).

3.9.2 Electrochemical genosensors (sandwich format) for the detection of bacterial pathogens in liquid biopsies

A sandwich assay-based DNA biosensor was developed for the detection of *Mycobacterium* sp. This involved sandwiching the target genomic DNA between a specific probe immobilized on an indium tin oxide (ITO) electrode modified with a (3-aminopropyl) trimethoxysilane SAM and detector probe/ alkaline phosphatase (AP)-dually labeled AuNPs for signal amplification (for references, Campuzano et al., 2017d). In another DNA-based sensing protocol, *Mycobacterium tuberculosis* was detected utilizing MBs and chronoamperometric detection at SPCEs with asymmetric helicase-dependent DNA amplification (aHDA). The aHDA method appeared to be a good alternative to PCR not requiring thermocycler allowing it to get easily adapted to the range of settings. This integrated strategy gave good results for the detection of *M. tuberculosis* from sputum, urine, and pleural fluid samples.

3.9.3 Electrochemical immunosensing of bacterial pathogens in liquid biopsies

In a study, electrodes were modified with tetrahedral surface DNA nanostructures for electrochemical immunosensing of *S. pneumoniae*. Hybridization chain reaction (HCR) was carried out at constant temperature without enzymes in conjunction with electrochemical sensors for the detection of bacterial infections. Lobo-Castañón's group has pioneered on-surface-helicase dependent amplification protocol for the quantification of genomic DNA isolated from *Salmonella* (Barreda-García et al., 2018). Electrochemical immunosensors can detect the whole bacteria (*Streptococcus pseudopneumoniae*, *Escherichia coli*, and *Vibrio cholerae*) specific surface proteins (PspA peptide), or specific bacterial genes. A disposable amperometric magneto-immunosensor was developed using Protein A modified with magnetic microbeads and gold screen-printed electrodes (Au/SPE) for the selective determination of *Streptococcus pneumoniae*. A study led by Gayathri et al. (2016) has utilized a glassy carbon electrode (GCE) scaffold modified with thionine dye immobilized chitosan/functionalized-MWCNT composite (GCE/f-MWCNT-Chit@Th) for the detection of uropathogenic *Escherichia coli* (UPEC). A sandwich electrochemical immunosensing protocol was devised for the rapid detection of pneumococcal surface protein A (PspA) peptide. This involved immobilizing the captured antibody onto a gold electrode nanostructured with a DNA tetrahedron (DNATH) and using the same antibody conjugated with ferrocene (FeC-Ab) as a detector. Li et al. (2017) have utilized multifunctional nano-conjugates and subsequent signal amplification for sensitive and simultaneous detection of *Escherichia coli* O157:H7 and *Vibrio cholerae* O1 (for references, Campuzano et al., 2017d).

Wu et al. (2010) have developed a sandwich hybridization-based approach for the detection of *Escherichia coli* 16S ribosomal RNA (rRNA). The study utilized a biosensor having a 16-Au electrode array synthesized by photolithography modified with a ternary self-assembled monolayer of a thiolated specific DNA capture probe, mercaptohexanol (MCH), and dithiothreitol (DTT). Liu et al. (2011) have developed a rapid antimicrobial susceptibility test (AST) combining the phenotypic culture of bacteria in physiological samples and electrochemical genotypic molecular marker detection by targeting bacterial 16S rRNA. This study has utilized alternating current (AC) electrokinetic fluid motion and Joule heating-induced temperature elevation for enhancing sensor signal and minimizing the matrix effect. Optimized electrokinetic conditions have facilitated direct detection of bacterial pathogens in blood culture without prior purification. The same group has developed a multiplexed electrochemical biosensor for the diagnosis of infections in the bloodstream by the detection of species-specific 16S RNA sequences of bacteria such as *Staphylococcus aureus*, *Escherichia coli*, *Pseudomonas aeruginosa*, and *Proteus mirabilis* in physiological samples without pre-amplification step. A disposable amperometric DNA sensor was developed utilizing MBs and asymmetric polymerase chain reaction (aPCR) targeting a specific region of the pneumococcal *lytA* gene for the detection of *Streptococcus pneumoniae*. This protocol has facilitated distinction between closely related species of *Streptococus*, such as *S. pneumonia* and *S. mitis*. This has also enabled distinction between blood and urine samples. In a study, bisbenzimidazole trihydrochloride (intercalator) was mixed with amplicons obtained from direct PCR to detect *Porphyromonas gingivalis*, a bacterium causing periodontal diseasc (for references, Campuzano et al., 2017d).

3.9.4 Electrochemical immunosensing of invertebrate pathogens in liquid biopsies

An impedimetric DNA-based sensor was developed for determining the variable concentrations of *Leishmania infantum* from canine serum samples. The study involved immobilization of a specific DNA probe onto a gold electrode previously coated with a layer of 3-mercaptopropyltrimethoxysilane on a polyaniline matrix containing AuNPs (PANIAuNPs). Isothermal recombinase polymerase amplification (RPA) with double-labeled amplicons (MBs and AuNPs) along with chronoamperometric detection at SPCEs was utilized for the detection of *Leishmania* DNA (for references, Campuzano et al., 2017d) (Fig. 9).

3.10. Lab-on-chip and telemedicine

Biosensors in this category include electrochemical-based, optical-based, micromechanical-based transduction, and others. It has enabled the detection of many pathogens such as Ebstein-Barr Virus (EBV), human immunodeficiency virus (HIV), *Salmonella typhimurium*, H5N1 influenza, and many more. A smart phone-based POC sensor for zika, dengue, chikungunya, etc. was also developed. A smartphone devices equipped with screen display, camera for signal capturing, memory for data storage, connectivity modes (Bluetooth, Wi-Fi, NFC, etc.), and GPS are well suited for POC-based diagnostic applications. A GPS can track global pandemics using smartphone devices. Microphones in a smartphone can perform spirometry for evaluating lung capacity by blow sound measurement (for references Lafleur et al., 2016; Roda et al., 2017). The need for these kinds of biosensors arrived due to fluidic assays requiring sample transport. It is time-saving and uses a small sample volume.

Biosensors find interesting applications in telemedicine, owing to the features like portability, user-friendly, and economical. These can be applied for the

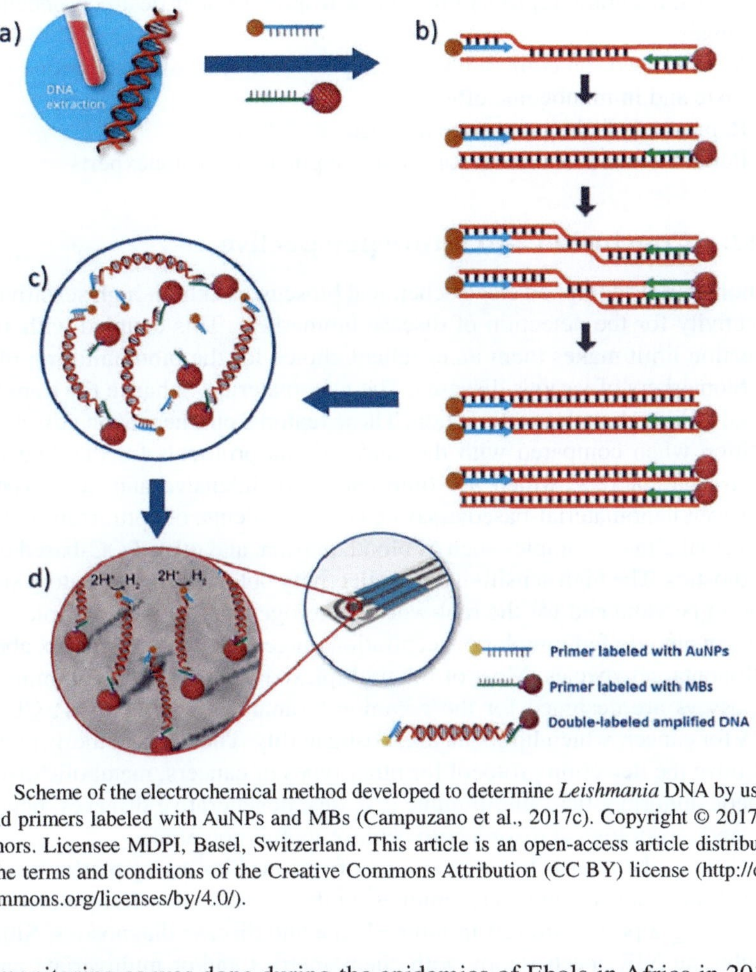

FIG. 9 Scheme of the electrochemical method developed to determine *Leishmania* DNA by using RPA and primers labeled with AuNPs and MBs (Campuzano et al., 2017c).

daily monitoring as was done during the epidemics of Ebola in Africa in 2014, SARS in 2003, and pandemic influenza in 2009 (Dziąbowska et al., 2017).

3.11. Optimal electrochemical biosensor characteristics (Gao and Lu, 2020; Huang et al., 2021; Zhang et al., 2020; Zuo et al., 2020)

An ideal biosensor is required to have the following attributes for its wider applicability in diverse real-world settings for the onsite, simple, rapid, sensitive, selective, and multiplexed determination of target analytes.

- Stability of biocatalyst under natural conditions
- Reaction independent of the physical changes such as temperature, pH, and agitation

- Linear, accurate, reproducible sensor response under desired concentration ranges
- Miniaturized, biocompatible sensor probe for clinical applications without toxic and immunogenic effects
- Rapid analysis in a real-time manner
- Portable, and economical for on-site application by nonexperts

3.12. Conclusion and future perspective

Nanomaterials equipped electrochemical biosensors exhibit high sensitivity and selectivity for the detection of disease biomarkers. This coupled with a lower detection limit makes them an excellent choice for the biomonitoring of clinical biomarkers of various diseases. The nanomaterials enhance the transducing signal of ligand-analyte interaction. These features put them at an advantageous position when compared with the conventional protocols for the detection of disease biomarkers, which are time and labor intensive and also expensive. However, nanomaterial-based assays face the challenge of optimization for real-life actual clinical samples such as blood, plasma, and other POC-based clinical diagnostics. The high sensitivity and selectivity obtained for laboratory samples need to be validated for the real-world, heterogeneous clinical samples, which come in pico- to femtomolar concentration range with complex matrix abundant with nontarget species. Most of the multiplexed electrochemical immunosensing assays are designed for the common biomarkers such as AFP, CEA, and PSA for cancer, which limits its diagnostic ability. There is an emergent need to optimize the detection protocol for other types of cancers, metabolic disorders, genetic abnormalities, autoimmune diseases, degenerative diseases, neurological, and physiological disorders to achieve its full potential.

Need of the hour is the conjugation of various biosensing platforms such as electrochemical and optical techniques for the amplification-free detection strategy having a proven impact in biomedicine and disease diagnostics. Similarly, combining NIR spectroscopy with chemometrics and/or multivariate analyses has proven to be a powerful, nondestructive, rapid, and simple analytical technique that may be employed for the biomonitoring of disease biomarkers.

References

Ahmed, A., Rushworth, J.V., Hirst, N.A., Millner, P.A., 2014. Biosensors for whole-cell bacterial detection. Clin. Microbiol. Rev. 27, 631–646.

Barreda-García, S., Miranda-Castro, R., de-los-Santos-Álvarez, N., Miranda-Ordieres, A.J., Lobo-Castañón, M.J., 2018. Helicase-dependent isothermal amplification: a novel tool in the development of molecular-based analytical systems for rapid pathogen detection. Anal. Bioanal. Chem. 410 (3), 679–693.

Benvidi, A., Dehghani Firouzabadi, A., Dehghan Tezerjani, M., Moshtaghiun, S.M., Mazloum-Ardakani, M., Ansarin, A., 2015. A highly sensitive and selective electrochemical DNA biosensor to diagnose breast cancer. J. Electroanal. Chem. 750, 57–64.

Bode, A.M., Dong, Z., 2004. Post-translational modification of p53 in tumorigenesis. Nat. Rev. Cancer 4, 793–805.

Campuzano, S., Pedrero, M., Nikoleli, G.P., Pingarrón, J.M., Nikolelis, D.P., 2017a. Hybrid 2D-nanomaterials-based electrochemical immunosensing strategies for clinical biomarkers determination. Biosens. Bioelectron. 89, 269–279.

Campuzano, S., Pedrero, M., Pingarrón, J.M., 2017b. Non-invasive breast cancer diagnosis through electrochemical biosensing at different molecular levels. Sensors 17, 1993.

Campuzano, S., Yáñez-Sedeño, P., Pingarrón, J.M., 2017c. Electrochemical genosensing of circulating biomarkers. Sensors 17, 866.

Campuzano, S., Yáñez-Sedeño, P., Pingarrón, J.M., 2017d. Molecular biosensors for electrochemical detection of infectious pathogens in liquid biopsies: current trends and challenges. Sensors 17, 2533.

Castañeda, M.T., Merkoçi, A., Pumera, M., Alegret, S., 2007. Electrochemical genosensors for biomedical applications based on gold nanoparticles. Biosens. Bioelectron. 22, 1961–1967.

Chandra, N., Liu, Y., Liu, J.X., Frängsmyr, L., Wu, N., Silva, L.M., Lindström, M., Chai, W., Pedrosa Domellöf, F., Feizi, T., Arnberg, N., 2019. Sulfated glycosaminoglycans as viral decoy receptors for human adenovirus type 37. Viruses 11 (3). pii: E247.

Chen, J., Zhang, J., Guo, Y., Li, J., Fu, F., Yang, H.H., Chen, G., 2011. An ultrasensitive electrochemical biosensor for detection of DNA species related to oral cancer based on nuclease-assisted target recycling and amplification of DNAzyme. Chem. Commun. 47, 8004–8006.

Chikkaveeraiah, B.V., Mania, V., Patel, V., Gutkind, J.S., Rusling, J.F., 2011. Microfluidic electrochemical immunoarray for ultrasensitive detection of two cancer biomarker proteins in serum. Biosens. Bioelectron. 26, 4477–4483.

Collinge, J., 2011. Prion diseases of humans and animals: their causes and molecular basis. Annu. Rev. Neurosci. 24, 519–550.

Costa Rama, E., Costa-García, A., 2016. Screen-printed electrochemical immunosensors for the detection of cancer and cardiovascular biomarkers. Electroanalysis 28, 1700–1715.

Cui, Z., Wu, D., Zhang, Y., Ma, H., Li, H., Du, B., Wei, Q., Ju, H., 2014. Ultrasensitive electrochemical immunosensors for multiplexed determination using mesoporous platinum nanoparticles as nonenzymatic labels. Anal. Chim. Acta 807, 44–50.

Darwish, N.T., Alrawi, A.H., Sekaran, S.D., Alias, Y., Khor, S.M., 2016. Electrochemical immunosensor based on antibody-nanoparticle hybrid for specific detection of the dengue virus NS1 biomarker. J. Electrochem. Soc. 163, B19–B25.

Dziąbowska, K., Czaczyk, E., Nidzworski, D., 2017. Application of Electrochemical Methods in Biosensing Technologies. Intech open.

Eletxigerra, U., Martinez-Perdiguero, J., Merino, S., Barderas, R., Ruiz-Valdepeñas Montiel, V., Villalonga, R., Pingarrón, J.M., Campuzano, S., 2016. Estrogen receptor α in serum, cell lysates, and breast cancer cell using an amperometric magnetoimmunosensing platform. Sens. Bio Sens. Res. 7, 71–76.

Emami, M., Shamsipur, M., Saber, R., Irajirad, R., 2014. An electrochemical immunosensor for detection of a breast cancer biomarker based on antiHER2-iron oxide nanoparticle bioconjugates. Analyst 139, 2858–2866.

Ensafi, A.A., Nasr-Esfahani, P., Heydari-Bafrooei, E., Rezaei, B., 2014. Redox targeting of DNA anchored to MWCNTs and TiO2 nanoparticles dispersed in poly dialyldimethylammonium chloride and chitosan. Colloids Surf. B Biointerfaces 121, 99–105.

Esteban-Fernández de Ávila, B., Araque, E., Campuzano, S., Pedrero, M., Dalkiran, B., Barderas, R., Villalonga, R., Kiliç, E., Pingarrón, J.M., 2015. Dual functional graphene derivative-based electrochemical platforms for detection of the TP53 gene with single nucleotide polymorphism selectivity in biological samples. Anal. Chem. 87, 2290–2298.

Farzin, L., Shamsipur, M., Samandari, L., Sheibani, S., 2018. Recent advances in designing nanomaterial based biointerfaces for electrochemical biosensing cardiovascular biomarkers. J. Pharm. Biomed. Anal. 161, 344–376.

Gao, N., Lu, K., 2020. An underwater metamaterial for broadband acoustic absorption at low frequency. Appl. Acoust. 169, 107500.

Gayathri, C.H., Mayuri, P., Sankaran, K., Kumar, A.S., 2016. An electrochemical immunosensor for efficient detection of uropathogenic E. coli based on thionine dye immobilized chitosan/functionalized-MWCNT modified electrode. Biosens. Bioelectron. 82, 71–77.

Ge, X., Zhang, A., Lin, Y., Du, D., 2016. Simultaneous immunoassay of phosphorylated proteins based on apoferritin templated metallic phosphates as voltammetrically distinguishable signal reporters. Biosens. Bioelectron. 80, 201–207.

Gupta, R.K., Pandya, R., Sieffert, T., Meyyappan, M., Koehne, J.E., 2016. Multiplexed electrochemical immunosensor for label-free detection of cardiac markers using a carbon nanofiber array chip. J. Electroanal. Chem. 773, 53–62.

Huang, Y., Wang, J., Wang, F., He, B., 2021. Event-triggered adaptive finite-time tracking control for full state constraints nonlinear systems with parameter uncertainties and given transient performance. ISA Trans. 108, 131–143.

Hwang, H., Ryu, M., Park, C.Y., Ahn, J., Park, H.G., Choi, C., Ha, S.D., Park, T.J., Park, J.P., 2017. High sensitive and selective electrochemical biosensor: label-free detection of human norovirus using affinity peptide as molecular binder. Biosens. Bioelectron. 87, 164–170.

Karapetis, S., Nikolelis, D., Hianik, T., 2018. Label-free and redox markers-based electrochemical Aptasensors for aflatoxin M1 detection. Sensors 18, 4218.

Lafleur, J.P., Jönsson, A., Senkbeil, S., 2016. Recent advances in lab-on-a-chip for biosensing applications. Biosens. Bioelectron. 76, 213–233.

Li, L., Feng, D., Zhang, Y., 2016. Simultaneous detection of two tumor markers using silver and gold nanoparticles decorated carbon nanospheres as labels. Anal. Biochem. 505, 59–65.

Li, Y., Xiong, Y., Fang, L., Jiang, L., Huang, H., Deng, H., Liang, W., Zheng, J., 2017. An electrochemical strategy using multifunctional nanoconjugates for efficient simultaneous detection of *Escherichia coli* O157: H7 and *Vibrio cholerae* O1. Theranostics 7 (4), 935–944.

Lin, M., Song, P., Zhou, G., Zuo, X., Aldalbahi, A., Lou, X., Shi, J., Fan, C., 2016. Electrochemical detection of nucleic acids, proteins, small molecules and cells using a DNA-nanostructure-based universal biosensing platform. Nat. Protoc. 11, 1244–1263.

Liu, B.Q., Zhang, B., Cui, Y.L., 2011. Multifunctional gold-silica nanostructures for ultrasensitive electrochemical immunoassay of streptomycin residues. ACS Appl. Mater. Interfaces 3, 4668–4676.

Low, K.F., Rijiravanich, P., Singh, K.K.B., Surareungchai, W., Yean, C.Y., 2015. An electrochemical genosensing assay based on magnetic beads and gold nanoparticle-loaded latex microspheres for vibrio cholerae detection. J. Biomed. Nanotechnol. 11, 702–710.

Majdinasab, M., Yaqub, M., Rahim, A., Catanante, G., Hayat, A., Marty, J.L., 2017. An overview on recent progress in electrochemical biosensors for antimicrobial drug residues in animal-derived food. Sensors 17, 1947.

Miodek, A., Castillo, G., Hianik, T., Korri-Youssoufi, H., 2013. Electrochemical aptasensor of human cellular prion based on multiwalled carbon nanotubes modified with dendrimers: a platform for connecting redox markers and aptamers. Anal. Chem. 85, 7704–7712.

Moreno-Guzmán, M., González-Cortés, A., Yáñez-Sedeño, P., Pingarrón, J.M., 2012. Multiplexed ultrasensitive determination of adreno-corticotropinand cortisol hormones at a dual electrochemical immunosensor. Electroanalysis 24, 1100–1108.

Oliveira, N., Souza, E., Ferreira, D., Zanforlin, D., Bezerra, W., Borba, M.A., Arruda, M., Lopes, K., Nascimento, G., Martins, D., Cordeiro, M., Lima-Filho, J., 2015. A sensitive and selective label-free electrochemical DNA biosensor for the detection of specific dengue virus serotype 3 sequences. Sensors 15, 15562–15577.

Pan, Y., Sonn, G.A., Sin, M.L.Y., Mach, K.E., Shih, M.C., Gau, V., Wong, P.K., Liao, J.C., 2010. Electrochemical immunosensor detection of urinary lactoferrin in clinical samples for urinary tract infection diagnosis. Biosens. Bioelectron. 26, 649–654.

Peng, D., Liang, R.P., Huang, H., Qiu, J.D., 2016. Electrochemical immunosensor for carcinoembryonic antigen based on signal amplification strategy of graphene and Fe_3O_4/Au NPs. J. Electroanal. Chem. 761, 112–117.

Pinto, A.M., Gonçalves, I.C., Magalhães, F.D., 2013. Graphene-based materials biocompatibility: a review. Colloids Surf. B Biointerfaces 111, 188–202.

Roda, A., Calabretta, M.M., Calab, D., 2017. Smartphone-based biosensors for bioanalytics: a critical review. Comp. Anal. Chem. 77, 237–286.

Salimian, R., Kékedy-Nagy, L., Ferapontova, E.E., 2017. Specific picomolar detection of a breast cancer biomarker HER-2/neu protein in serum: electrocatalytically amplified electroanalysis by the aptamer/PEG-modified electrode. Chem. Electro. Chem. 4, 872–879.

Sánchez-Tirado, R., Salvo, C., González-Cortés, A., Yáñez-Sedeño, P., Langa, F., Pingarrón, J.M., 2017. Electrochemical immunosensor for simultaneous determination of interleukin-1 beta and tumor necrosis factor alpha in serum and saliva using dual screen printed electrodes modified with functionalized double-walled carbon nanotubes. Anal. Chim. Acta 959, 66–73.

Shuai, H.L., Huang, K.J., Xing, L.L., Chen, Y.X., 2016. Ultrasensitive electrochemical sensing platform for microRNA based on tungsten oxide-graphene composites coupling with catalyzed hairpin assembly target recycling and enzyme signal amplification. Biosens. Bioelectron. 86, 337–345.

Singh, P., 2014. Surface Plasmon Resonance. Nova Science Publishers, New York, USA.

Singh, P., 2016a. SPR Biosensors: historical perspectives and current challenges. Sensors Actuators B Chem. 229, 110–130.

Singh, P., 2016b. In: Giddes, C.D. (Ed.), LSPR Biosensing: Recent Advances and Approaches in Reviews in plasmonics. 2017. Springer International Publishing, AG, pp. 211–238.

Singh, P., 2017a. G protein coupled receptors: a potential candidate for drug discovery. Int. J. Pharmacol. 13, 846–863 (Special Issue).

Singh, P., 2017b. Surface Plasmon Resonance: A Boon for Viral Diagnostics in Reference Module in Life Sciences. Elsevier, Netherlands.

Sun, G., Zhang, L., Zhang, Y., Yang, H., Ma, C., Ge, S., Yan, M., Yu, J., Song, X., 2015. Multiplexed enzyme-free electrochemical immunosensor based on ZnO nanorods modified reduced graphene oxide-paper electrode and silver deposition-induced signal amplification strategy. Biosens. Bioelectron. 71, 30–36.

Verma, P., Maire, P., Novak, P., 2011. Concatenation of electrochemical grafting with chemical or electrochemical modification for preparing electrodes with specific surface functionality. Electrochim. Acta 56, 3555–3561.

Wang, K., He, M.Q., Zhai, F.H., He, R.H., Yu, Y.L., 2017. A novel electrochemical biosensor based on polyadenine modified aptamer for label-free and ultrasensitive detection of human breast cancer cells. Talanta 166, 87–92.

Wu, J., Campuzano, S., Halford, C., Haake, D.A., Wang, J., 2010. Ternary surface monolayers for ultrasensitive (zeptomole) amperometric detection of nucleic acid hybridization without signal amplification. Anal. Chem. 82 (21), 8830–8837.

Wu, L., Ji, H., Sun, H., Ding, C., Ren, J., Qu, X., 2016. Label-free ratiometric electrochemical detection of the mutated apolipoprotein E gene associated with Alzheimer's disease. Chem. Commun. 52, 12080–12083.

Wu, Q., Zhang, Y., Yang, Q., Yuan, N., Zhang, W., 2019. Review of electrochemical DNA biosensors for detecting food borne pathogens. Sensors 19, 4916.

Xu, H., Zhu, X., Wang, J., Lin, Z., Chen, G., 2019. Electrochemiluminescent functional nucleic acids-based sensors for food analysis. Luminescence 34, 508–515.

Yáñez-Sedeño, P., Campuzano, S., Pingarrón, J.M., 2019. Pushing the limits of electrochemistry toward challenging applications in clinical diagnosis, prognosis, and therapeutic action. Chem. Commun. 55 (18), 2563–2592.

Yang, T., Gao, Y., Liu, Z., Xu, J., Lu, L., Yu, Y., 2017. Three-dimensional gold nanoparticles/prussian blue-poly(3,4-ethylenedioxythiophene) nanocomposite as novel redox matrix for label-free electrochemical immunoassay of carcinoembryonic antigen. Sensors Actuators B Chem. 239, 76–84.

Yang, Z.-H., Zhuo, Y., Chai, Y.-Q., Yuan, R., 2014. High throughput immunosenor based on multi-label strategy and a novel array electrode. Sci. Rep. 4, 4747.

Zhang, R., Jiang, T., Li, F., Li, G., Chen, H., Li, X., 2020. Coordinated bidding strategy of wind farms and power-to-gas facilities using a cooperative game approach. IEEE Trans. Sustainable Energy 11, 2545–2555.

Zhang, G.J., Luo, Z.H.H., Huang, M.J., Ang, J.J., Kang, T.G., Ji, H., 2011. An integrated chip for rapid, sensitive, and multiplexed detection of cardiac biomarkers from fingerprick blood. Biosens. Bioelectron. 28, 459–463.

Zhao, C., Liu, X., 2016. A portable paper-based microfluidic platform for multiplexed electrochemical detection of human immunodeficiency virus and hepatitis C virus antibodies in serum. Biomicrofluidics 10, 024119.

Zhao, S., Yang, W., Lai, R.Y., 2011. A folding-based electrochemical aptasensor for detection of vascular endothelial growth factor in human whole blood. Biosens. Bioelectron. 26, 2442–2447.

Zheng, L., Niu, X., Zhao, J., Liu, T., Liu, Y., Yang, Y., 2016. A label-free immunosensor for CEA based on Pd-Ir bimetallic nanoparticles. J. Nanosci. Nanotechnol. 16, 5984–5990.

Zhou, S., Wang, Y., Zhu, J.J., 2016. Simultaneous detection of tumor cell apoptosis regulators Bcl-2 and Bax through a dual-signal-marked electrochemical immunosensor. ACS Appl. Mater. Interfaces 8, 7674–7682.

Zuo, X., Dong, M., Gao, F., Tian, S., 2020. The modeling of the electric heating and cooling system of the integrated energy system in the coastal area. J. Coast. Res. 103, 1022–1029.

Chapter 4

Electrochemical nano-biosensors: Environmental biomonitoring

4.1. Introduction

Biological monitoring can determine total chemical exposure from multiple sources such as air, soil, water, and food residues taken through inhalation, oral or dermal routes. Biomonitoring can quantify the internal dose of a given chemical or metabolite on a particular tissue or the whole body. This can be utilized for risk assessment purposes. However, the detection and quantification of the chemicals, toxicant or a metabolite require the integration of numerous operating factors such as pharmacokinetics, identification of biomarkers in a complex biological matrix, validation of sensing platform (including a limit of detection, the limit of quantification, linear performance, reproducibility, matrix effects, etc.), its comparison with other well established analytical techniques and platforms (mass spectrometry and ELISA), and its on-site deployment. Further, in vivo animal, and in vitro human validation is required to evaluate the parameters like selectivity, sensitivity, and reproducibility of the sensing platforms. These requirements can be addressed by a miniaturized, portable biosensing device that can analyze multiple biomarkers in an inexpensive, rapid, and on-site manner.

This chapter highlights the advances in the development of electrochemical sensors with emphasis on the fabrication of nanomaterial-based miniaturized, portable, sensitive, selective, and flexible electrochemical sensors, where nanomaterials act as a signal-transducer or as an electroactive species for the indirect detection of an analyte. The chapter aims to critically consider the recently published innovations in this area and highlight the need to include more realistic assays in future research in order to make these advances suitable for on-site analysis.

The major aspects of the entire chapter incorporate the following: (i) electrochemical biosensors, affinity biosensors and its major types based on transducers, (ii) nanomaterials engineering for the advancements of electrochemical sensors, (iii) biomonitoring of organic-inorganic pollutants, (iv) biomonitoring

Electrochemical Biosensors. https://doi.org/10.1016/B978-0-323-90632-6.00002-9

of water-borne pathogens, (v) biomonitoring of agricultural pollutants, (vi) role of nanotechnology in remediation efforts, (vii) risk assessment and management of nanomaterials, and (viii) recent advances and future trends.

4.2. Electrochemical biosensors

Electrochemical biosensors measure the charges across an electrode. An amperometric detector measures the current when an electroactive solute contacts a working electrode held at a fixed potential with respect to a reference electrode while a voltametric detector measures the current that develops as a function of variable potential. Both amperometric and voltametric detectors quantify electrode current as a function of time. However, voltammetry provides characteristics current response for reversible reaction by pulsating applied potential between high and low values. When the potential is applied in a staircase fashion, a waveform is produced as a function of applied potential per unit time, the technique is known as square-wave-voltammetry (SWV) (Barry et al., 2009). Classical amperometric and voltametric electrochemical sensors offer low detection limits in the range of 10^{-7} to 10^{-8} (Bard and Faulkner, 1980). Most of the analytes are electroactive, therefore, electrochemical sensing techniques are advantageous over atomic/mass-spectroscopic techniques (Fig. 1).

Based on the electrical signal applied to the electrochemical cell, electrochemical techniques can be **potentiostatic** (controlled potential is applied to the electrochemical cell and current is measured); **galvanostatic** (current is applied and potential is measured); **potentiometric** (cell potential is measured under the condition of near-zero current); **impedance** (potential is applied to the cell and current response is measured and analyzed so that impedance or complex resistance is obtained) (Bard and Faulkner, 2001) (Fig. 2).

Electrochemical techniques are based on liquid samples using flow injection analysis (FIA) or sequential injection analysis (SIA) using amperometric detection. This enables the continuous-flow detection of toxic compounds in environmental or biological samples in a real-time manner. The technique utilizes volumes in microliters (μL) for the high-throughput analysis of multiple

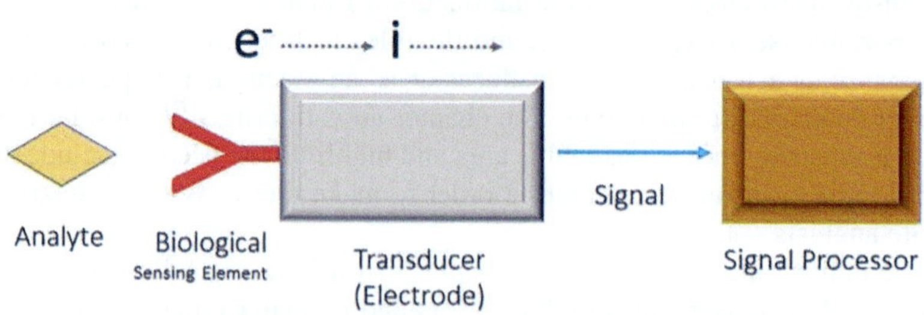

FIG. 1 Schematic illustration of mechanism/principle of the electrochemical biosensor (Mishra et al., 2018). *Copyright © 2018 by the authors. Licensee MDPI, Basel, Switzerland. This article is an open-access article distributed under the terms and conditions of the Creative Commons Attribution (CC BY) license (http://creativecommons.org/licenses/by/4.0/).*

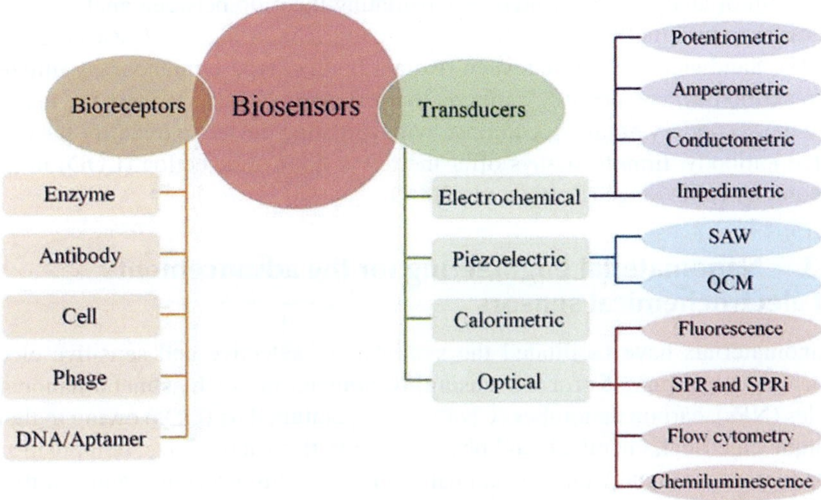

FIG. 2 Components and classification of the biosensor (Wu et al., 2019). *Copyright © 2019 by the authors. Licensee MDPI, Basel, Switzerland. This article is an open-access article distributed under the terms and conditions of the Creative Commons Attribution (CC BY) license (http://creativecommons.org/licenses/by/4.0/).*

samples (Wang and Li, 1990; Liu et al., 2005; Liu and Lin, 2005). However, despite its tremendous potential in the field of biomonitoring or biosensing, the application of electrochemical techniques is limited owing to the requirement of electrodes with unique electrical and selective properties (Barry et al., 2009).

4.2.1 Electrochemical affinity biosensors

Electrochemical affinity biosensors with lower detection limit, higher selectivity, and specificity for the target, portability, miniaturization, and ability to conjugate with other bioanalytical techniques are key for detecting the environmental contaminants or toxicants. This is further enhanced by integrating nanomaterial engineering with interface biochemistry. The electrochemical biosensors equipped with nanomaterials of carbon and metal nanocomposites lead to improvisation of transduction signal enabling the lower detection limit of femtomolar concentration (Loyprasert et al., 2008; Ruiyi et al., 2013).

Electrochemical immunosensors are based on the affinity between the antibody and a specific antigen. Immunoassays are of two types, viz. competitive active site assay, and sandwich assay. Usually, the antibody or analyte of interest are immobilized on a membrane, magnetic bead, or electrochemical transducer, and the analyte is measured via a signal derived from a conjugated tag that is attached to either the antibody or analyte. Competitive immunoassay involves the competition between the analyte in the matrix and the enzyme/nanoparticle labeled analog analyte for the binding site on the antibody. In this assay, the quantum of analyte is measured by calculating the ratio between analog signal to maximum signal expected from the analog (Lin et al., 2007; Zacco et al., 2007). Sandwich (analyte between two antibodies) type immunoassay utilizes two antibodies, viz. primary antibody that captures the analyte from the matrix and the secondary antibody equipped with signaling device to quantify the captured antibody. Immunoassays offer the lower limit of detection (LOD) in the picomolar (pM) range (Barry et al., 2009).

4.3. Nanomaterial engineering for the advancement of electrochemical sensors

Nanomaterials have facilitated the synthesis of selective and sensitive electrochemical techniques for biosensing. Nanomaterials such as metal nanoparticles (NPs), carbon nanotubes (CNTs), and quantum dots (QDs) owing to their unique electrical, chemical, and physical properties such as size, composition, conductivity, magnetism, mechanical strength, light-absorbing, and emitting properties are used as transducers or electroactive tags for the detection of analytes (Barry et al., 2009). CNTs and metal NPs are used to modify electrode surfaces to improve the electrochemical responses (Chen et al., 2007). Some NPs are used as electroactive reporters for the indirect detection of analyte or signal amplification (Authier et al., 2001; Cui et al., 2007; Wu et al., 2007; Liu et al., 2007). Polymer nanowires and silicon nanowires are used as field-effect transistors (FET) (Patolsky et al., 2006). CNTs (graphite carbon) and carbon nanofiber possess exciting properties such as conductivity, surface areas, chemical functionalities, and biocompatibility. CNTs can enhance direct electron transfer of some biomolecules such as cytochrome *c*, catalase, and nicotinamide adenine dinucleotide (NAD) due to the unique electronic structure, electrical conductivity, and high surface area available on carbon nanotubes for redox reactions (Lin et al., 2005; Liu and Lin, 2005; Wang et al., 2006) (Fig. 3).

Nanoparticles such as colloidal metals (gold or silver), inorganic crystals (e.g., quantum dots), and silica in conjunction with biomolecules (DNA, protein, antibody) are used for studying biomolecular interactions (Guo and Wang, 2007). Nanoparticle-based composites (nanocomposites) are increasingly used for designing electrodes in electrochemical biosensors for the detection of environmental pollutants. These nanoparticles have a positive effect on the conductivity enhancement, introducing high-density molecular layers, and signal enhancement, etc. (Campàs et al., 2012; Reverté et al., 2016). Metallic nanoparticles in an electrochemical biosensor improve chemical stability, biocompatibility, facilitate electron transfer between the analyte and electrode surface (Vogiazi et al., 2019). Carbon-based nanomaterials such as single-walled carbon nanotube (SWCNTs), multi-walled carbon nanotube (MWCNTs), single-walled carbon nanohorns (SWCNHs), carbon nanospheres (CNS) are used to modify electrodes and act as a label for signal

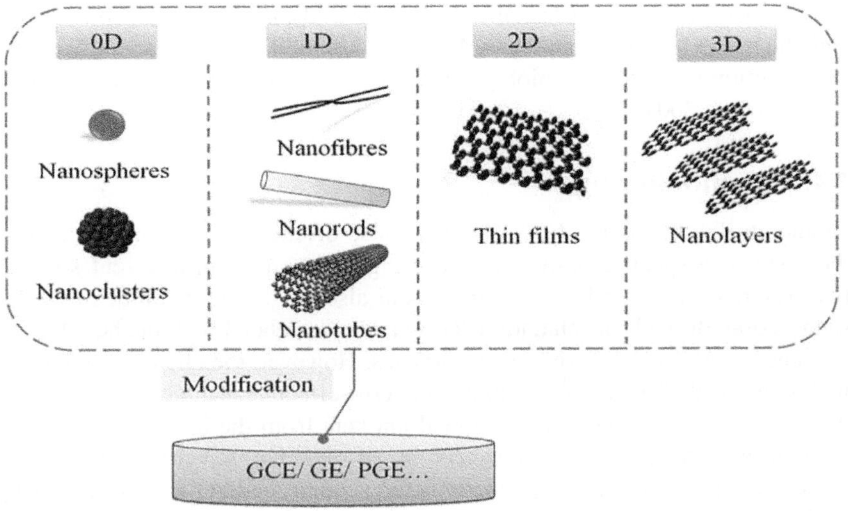

FIG. 3 Nanomaterials are commonly used in the modification of electrochemical DNA biosensors (Wu et al., 2019). *Copyright © 2019 by the authors. Licensee MDPI, Basel, Switzerland. This article is an open-access article distributed under the terms and conditions of the Creative Commons Attribution (CC BY) license (http://creativecommons.org/licenses/by/4.0/).*

amplification (Zhang et al., 2010). The use of 3D graphene has enabled the development of electrochemical biosensors with lower LOD (Han et al., 2013). Conjugation of carbon nanomaterials and metallic nanomaterials provide good electrode support, and an overall synergistic effect on the performance of the sensor such as high sensitivity, stability, and reproducibility (Ruiyi et al., 2013). Screen-printed electrodes (SPEs) usage in biosensors have led to the portability and miniaturization of electrodes resulting in the development of microelectrodes, ultramicroelectrodes, and nanoelectrodes (Vogiazi et al., 2019) (Fig. 3).

Advancements in printing technology and material science allow greater flexibility in designing selective, reliable, and economical electrodes for the detection of specific analytes. These advancements have resulted in screen-printed electrodes (SPE), integrated nanomaterials, and electrochemical immunoassays. These advanced systems are capable of the identification of chemicals, metabolites, proteins, metals, inorganic ions, organic compounds, and biomolecules in a variety of biological matrices (Barry et al., 2009).

Screen-printed electrodes (SPE) are synthesized by carbon-based materials (metals, complexing molecules, immobilized enzymes, or affinity agents such as antibodies) coated with doping materials (gold or silver atoms, or coated with gold, mercury, bismuth, or nickel metal) to enhance electron transfer or coated with selective agents to capture analytes of interest using SPE (photolithography). Enzyme-immobilized SPEs are used for the detection of enzyme activity, pesticide metabolites, phenolic compounds, heavy metals, cholesterol, and glucose in biological matrices. Antibody-immobilized SPEs are utilized for the detection of antigen in biological matrices (Badihi-Mossberg et al., 2007; Renedo et al., 2007).

4.4. Biomonitoring

Biomonitoring can quantify the internal dose of chemicals or metabolites in the body or a specific tissue by measuring exposure from different sources like water, soil, air, and food residues and also by different routes of intake such as oral, dermal' or inhalation. This can be co-related with the health consequences and for risk assessment purposes. However, the chemical exposure with biological effect requires simple and cost-effective techniques and assays that can detect and quantitate chemical markers from the biological fluids in a noninvasive manner. A nanomaterial-based sensitive, selective, flexible on-site electrochemical sensing platform is the need of the hour for the chemical-biomonitoring in complex matrices (urine, blood, saliva, and other body fluids). Ashley (2003) has reviewed the electrochemical sensors for the on-site monitoring of chemical contaminants such as inorganic gases and vapors (carbon monoxide, carbon dioxide, sulfur dioxide, and nitric oxide), volatile organic hydrocarbons, aldehydes and ketones, heavy metals, pesticides, and some persistent pollutants. Numerous sensing platforms have been developed for the monitoring of toxic metals or chemical xenobiotics in environment applications (Barry et al., 2009; Hanrahan et al., 2004; Sole and Alegret, 2001; Sadik and Van Emon, 1996).

4.5. Chemical contaminants in water

Electrochemical sensors are made up of organic-inorganic nanomaterials that show the ability to interact with different pollutants spread in the environment. These can be synthesized on large scale to fabricate chemical sensing devices. Electrochemical sensors based on metal/metal-oxide nanoparticles doped with conductive porous composites demonstrate enhanced sensitivity for the detection and quantification of water pollutants owing to innumerable accessible and functional sites provided by the interconnected porosity and the metal nanoparticles. The porous carbon composites having metallic nanoparticles synthesized via sol-gel techniques can be readily processed into porous hybrid materials like powders and thin films. These can be used in electrode fabrication based on screen-printing and lithography techniques. The electrochemical sensors based on metal nanoparticles (Bismuth and Gold) can be miniaturized on chip-sensing devices for the determination of heavy metal pollutants in water. The rapidly changing market requires the design of electrochemical sensors, which are fast, reliable, and accurate for the on-site analysis of a wide variety of target analytes (Niu et al., 2018).

4.5.1 Heavy metals

Heavy metal ions especially Lead (Pb), Mercury (Hg), Arsenic (As), Chromium (Cr), and Cadmium (Cd) are nonbiodegradable. These heavy metals contaminate most of the natural resources especially water. Water contaminated with heavy metals leads to serious health risks. It affects health management efforts as even the trace amount of exposure to these heavy metals is highly toxic and poses a serious threat to the health of both plants and animals including the human population. Therefore, assays and techniques are required that can detect these hazardous inorganic-organic chemicals even in the trace amount in a simple, rapid, low-cost, on-site, real-time manner.

4.5.2 Lead (Pb^{2+})

Ji et al. (2019) have designed a novel Pb^{2+} electrochemical DNAzyme sensor for the specific and ultrasensitive detection of lead ions (Pb^{2+}) in water. The architecture consists of MoS_2-AuPt nanomaterials and hemin/G-quadruplex DNAzyme serving as the electrocatalytic signal tag. The sensing platform has utilized graphene oxide (rGO-SnO_2)/gold nanoparticles (AuNPs) for conductivity enhancement for immobilizing more Pb^{2+}-specific DNAzyme. Hemin/G-quadruplex (G-rich DNA) enabled signal amplification while chronoamperometry was utilized to record signal.

4.5.3 Mercury (Hg^{2+})

Exposure to Mercury (Hg) is a major public health and safety concern in both acute and chronic cases of exposure. Young children, pregnant women, and lactating mothers are particularly more vulnerable. Hg is known to cause potent neurologic, renal, and developmental toxicities leading to a number of disorders and diseases, the prominent being "Mad hatter syndrome" and "Minamata disease" in addition to brain damage, kidney failure, neurological disorders, and chromosomal abnormalities. The effect of Hg particularly becomes more dangerous through the process of biomagnification where it gets more concentrated from the lower to the higher trophic level of the food chain. This feature has put mercury (Hg) high on ranking in the list prepared by the Agency for Toxic Substances and Disease Registry. According to the guidelines of the World Health Organization (WHO), the human can tolerate 30 nM of inorganic Hg in drinking water, while the United States Environmental Protection Agency (USEPA) recommends this limit to be 10 nM only (US EPA, 2001). The Potent route of exposure is through water and food contaminated with Hg from both man-made and natural sources. Therefore, assays and techniques are required that lead to simple and rapid detection of solvated aqueous Hg^{2+} in trace quantity for its effective management.

Traditional quantitative methods, such as atomic absorption/emission spectroscopy, inductively coupled plasma mass spectrometry, and cold vapor atomic fluorescence spectrometry (CVAFS) can detect and quantify heavy metals with high sensitivity. However, these techniques involve a cumbersome and complicated chemical process for the extraction of intended metal, and are expensive, time-intensive, and labor-intensive (for references Long et al., 2013). Different types of sensors for the detection of solvated aqueous Hg^{2+} in the solution are reported. However, recent efforts are focused on designing the sensors based on carbon nanotubes (CNTs) owing to their specificity and sensitivity towards the detection of Hg^{2+} ions in trace amounts. CNT-based sensors for Hg^{2+} detection is potentiometric, DNA-based fluorescence, surface plasmon resonance (SPR), colorimetric, and stripping voltammetric (SV) assays. These sensors have their inherent advantages and disadvantages in terms of ease of fabrication, the economy of the cost involved, sensitivity, selectivity, ease of operation, the limit of detection, etc., further requiring fabrication of next-generation improvised sensors crucial for the need-based effective management of Hg pollution (Pokhrel et al., 2017).

Somerset et al. (2011) have designed an electrochemical sensor employing screen-printed carbon electrodes (SPCEs) and carbon paste electrodes (CPEs) for the detection of trace metal ions in aqueous solutions such as Hg^{2+}, Pb^{2+}, Ni^{2+}, and Cd^{2+}. Designed architecture consists of polyaniline (PANI), or polyaniline-poly(2,2′-dithiodianiline) (PANI-PDTDA) coated SPCEs along with electroactive compounds containing CPEs for increased selectivity. The modified carbon paste electrodes (MCPEs) and SPCE/PANI-PDTDA sensors have shown that the concentration range of 1×10^{-9}M and 1×10^{-6}M; deposition potential of -400 mV; and accumulation time of 120 s was adequate for the detection of Hg^{2+}, Pb^{2+}, Ni^{2+}, and Cd^{2+} using differently constructed electrodes (Somerset et al., 2011).

Xu-Cheng et al. (2012) have devised an electrochemical detection protocol for Hg(II) utilizing electropolymerized ion imprinting poly(2-mercaptobenzothiazole) films at the surface of gold nanoparticles/single-walled carbon nanotube nanohybrids (AuNP/SWCNT) modified glassy carbon electrode (PMBT/AuNPs/SWCNTs/GCE). The designed sensor exhibited larger binding to functionalized capacity, larger affinity, faster binding kinetics, and higher selectivity to template Hg^{2+}. The differential pulse anodic stripping voltammetry (DPASV) response of the Hg^{2+}-imprinted PMBT/AuNPs/SWCNTs/GCE sensor to Hg^{2+} was found to be 3.7- and 10.5-fold higher than nonimprinted PMBT/AuNPs/SWCNTs/GCE and the imprinted PMBT/AuNPs/GCE, respectively. The limit of detection for Hg^{2+} was found below the value given by World Health Organization (WHO). The sensor showed high sensitivity, selectivity, and reproducibility. Overall the performance of Hg^{2+}-imprinted PMBT/AuNPs/SWCNTs/GCE towards Hg(II) was found to be superior to the existing electrodes for the detection of Hg^{2+} from the natural water samples (Xu-Cheng et al., 2012).

A structure-switching DNA evanescent wave fiberoptic biosensor was fabricated for the rapid on-site/in situ detection of Hg^{2+} from the real-world samples. The-DNA based sensor was designed to exploit the ability of certain metal ions to selectively bind nitrogenous bases in DNA resulting in the formation of metal-induced DNA duplexes. Hg^{2+} can coordinate with thymidine bases to form T-Hg^{2+}-T coordination complex (Clever et al., 2007; Willner and Zayats, 2007; Tang et al., 2010). The platform consists of a DNA probe immobilized onto the sensor surface to which fluorescence-labeled DNA containing T-T mismatch structure was added as an analyte. This has allowed the labeled DNA to bind with Hg^{2+} forming T-Hg^{2+}-T complex accompanied with the reduction in fluorescence signal. The evanescent wave fiberoptic biosensor exhibits the added advantage of sensitivity, speed, portability, and minimum sample manipulation. This sensing platform can be extended to other heavy metals by replacing T-Hg^{2+}-T complex with other specific structures that selectively binds to other analytes. For instance, Ag^+ interacts selectively and specifically with cytosine-cytosine (C-C) mismatches, while Pb^{2+} with Pb^{2+} dependent DNAzyme and Pb^{2+} stabilized G quadruplexes (Xiao et al., 2007; Ono et al., 2008; Li et al., 2010; Long et al., 2013).

A new rhodamine-based monomer (RD-CZ) was electropolymerized on an indium tin oxide (ITO) electrode P (RD-CZ). The resulting polymer film was utilized as a simple and novel multi-signal sensor platform exhibiting ion-selective potentiometric, colorimetric, and voltammetric responses in aqueous media. Compared to Cd^{2+}, Cu^{2+}, Zn^{2+}, and Fe^{3+} ions, P (RD-CZ) has demonstrated excellent selectivity for Hg^{2+} ions utilizing the potentiometric techniques. The P (RD-CZ) polymer film gives colored signals, i.e., orange to purple. Overall, a low-cost, point-of-care (POC) conductive polymer-based sensing platform was designed for the on-site detection of Hg^{2+} ions in aqueous media. This sensor demonstrated a linear relationship with respect to the analyte concentration in aqueous media and biological samples (Ayranci et al., 2017).

4.5.4 Arsenic (As^{3+}, As^{5+})

Similar to the inorganic mercury Hg(II), inorganic arsenic As(III), and As(V) also pose serious health hazards. Numerous health complications are reported among the people exposed to underground water contaminated with arsenic species. Various electrochemical methods especially stripping voltammetry (SV) are employed for the simple and rapid detection of arsenic. However, modified carbon paste electrode (MCPE) based sensors provide better detection owing to a better deposition with a linear and improved response (Zaib et al., 2015).

Recent years have seen a spurt in the design and development of novel nanoparticle-based sensors for the detection and quantitation of heavy metals particularly, Hg^{2+}, Pb^{2+}, Ni^{2+}, Cd^{2+}, Zn^{2+}, Cu^{2+}, Fe^{3+}, As^{3+}, and As^{5+} from natural and man-made samples in simple, rapid, reproducible, reliable, and economical manner. Various methodologies such as colorimetric, fluorometric, magnetic, and electrochemical output signals were employed allowing the detection limit within the pM and nM ranges. However, these protocols still lag behind the optimized and standardized cold vapor atomic absorption spectrometry and inductively coupled plasma techniques in terms of reliability and sensitivity. This is because the sample matrix largely influences the response of nanoparticle-based sensors. The designed sensors may fail in the real-world sample testing due to the negative incidence of the ionic strength and the presence of exchangeable ligands (Botasini et al., 2013).

4.6. Phenolic compounds

Phenols are a very common by-product of many industrial processes, which are extremely toxic and harmful. These are usually present in food, fresh water, and wastewater. Phenols and phenolic derivatives are released into the environment by a range of human activities, both industrial and domestic and are toxic to the health of the environment and humans even in the trace amount. The reliable detection of phenols and phenolic derivatives is of utmost concern for the health, sustainability, and management of the environment. Metal oxide nanocomposites show interesting applications in the field of nanomaterial engineering for the design and construction of sensors, especially for the on-site monitoring of air-water contamination and other toxic pollutants in the environment. Metal oxide nanocomposites show interesting features like dynamic surface area, elevated porosity, permeability, stability, quantum confinement consequence, mesoporous nature, and excellent absorption. The process of adsorption makes it appropriate for the reliable detection and quantification of environmental pollutants (for references Rahman et al., 2017).

2-Nitrophenol (2-NP) belongs to the nitro-aromatic group of chemicals. These are extensively used in the production of explosives, fungicides, herbicides, insecticides, pesticides, pharmaceuticals, petroleum products, pigments, synthetic dyes, rubber chemicals, textiles, and wood (Isichei and Okieimen, 2014; Zhu et al., 2014). 2-NP is a highly persistent and hazardous organic pollutant and exhibits very high toxicity and mutagenic activity (Anmar et al., 2007; Rebeiro et al., 2013). A nanocomposite (Ce_2O_3.CNT NCs) and carbon nanotube (CNT) based sensor having Ce_2O_3 nanoparticle was synthesized by a wet chemical method in an alkaline medium for the detection and quantification of nitrophenols. The nanocomposite (Ce_2O_3.CNT NCs) was further characterized by FTIR, UV/Vis, Field-Emission Scanning Electron Microscopy (FESEM), X-ray electron dispersive spectroscopy (XEDS), X-ray photoelectron spectroscopy (XPS), and powder X-ray diffraction (XRD). The thin layer of nanocomposite was deposited onto a flat glassy carbon electrode (GCE). The efficacy of the sensor was determined by the selective and sensitive identification of toxic 2-nitrophenol (2-NP) in real environmental samples (Hussain et al., 2016).

3-Methoxy Phenol poses a serious threat to the environment and human health and hence requires techniques and assays, which can detect 3-MP in a highly sensitive, selective, and economical manner. Various electrochemical methods such as spectrometry, high-pressure liquid chromatography (HPLC), etc. have been applied for the effective detection and quantification of phenols in environmental samples (Rahman et al., 2011, 2015). A nanocomposite (NC) based sensor was fabricated for the detection of 3-methoxyphenol (3-MP). The NC was synthesized using Iron oxide (Fe_3O_4) and carbon nanotube (CNT), (Fe_3O_4.CNT NCs) via a wet-chemical process in an alkaline medium. The nanocomposite (Fe_3O_4.CNT NCs) was characterized for optical, morphological, and structural features through FTIR, UV/Vis., FESEM, TEM, XEDS, XPS, and XRD respectively. A thin layer of NC was deposited on the flat GCE using a coating binding agent. The nanocomposite-based sensor has shown enhanced sensitivity, lower detection limit, linear dynamic range (LDR), and selectivity for 3-MP (Mohammed et al., 2017). Thus, validating the basic design and fabrication process (nanocomposites via wet-chemical progression) having an important application for the monitoring and regulation of toxic chemicals in the environment (Rahman et al., 2017).

A nanocomposite (NC) sensor containing chromium (III) oxide nanomaterials integrated with carbon nanotubes (Cr_2O_3-CNT NC) was synthesized using reducing agents in an alkaline medium for the sensitive, and reliable detection of toxic phenolic compounds in the environment. This was further characterized by ultraviolet-visible (UV/Vis.) spectroscopy, Fourier-transform infrared (FTIR) spectroscopy, energy-dispersive X-ray spectroscopy (XEDS), X-ray powder diffraction (XRD), X-ray photoelectron spectroscopy (XPS), and field-emission scanning electron microscopy (FESEM). The Cr_2O_3-CNT nanocomposite was deposited on a glassy carbon electrode (GCE) with Nafion binders enabling high selectivity and sensitivity towards 4-methoxyphenol (4MP) in phosphate buffer. The nanocomposite sensor has shown good selectivity, sensitivity, LOD, reliability, stability, reproducibility, and ability to get integrated with other bioanalytical tools (Rahman et al., 2017).

A nanocomposite using Nickel oxide nanoparticles integrated with carbon nanotube nanocomposites (NiO·CNT NCs) was synthesized in an alkaline medium by using facile wet-chemical routes. NiO·CNT NCs was characterized by Fourier transformed infrared (FT-IR), Ultra-violet visible (UV/Vis) spectroscopy, field-emission scanning electron microscopy (FESEM), X-ray energy dispersed spectroscopy (XEDS), X-ray photoelectron spectroscopy (XPS), and powder X-ray diffraction (XRD) for optical, morphological, and structural properties. A thin layer of nanocomposites was deposited on a flat glassy carbon electrode (GCE) to design **4-aminophenol (4-AP)** chemical sensor. The sensor has shown a good response with respect to the higher sensitivity, large dynamic range (LDR), the lower limit of detection (LOD), and long-term stability towards 4-AP. The wet-chemical method leads to the design of an ultrasensitive and selective chemical sensor for the detection of environmentally toxic substances. The 4-AP sensitive NiO·CNT NCs based sensor can also be used for the detection of toxic substances in biological samples and other point-of-care health diagnostic applications (Hussain et al., 2017).

4.7. Cyanotoxins

Rapidly growing algal blooms of cyanobacteria have become a potent environmental concern due to the odor, color, taste variations, and cyanotoxin production. These cyanotoxins can be hepatotoxic, neurotoxic, and dermatoxic. Chemically, these are cyclic heptapeptides like microcystins. It has over 200 variants such as cylindrospermopsin, anatoxin-a, and saxitoxin. These variants differ from each other in terms of methylation, epimerization, hydroxylation, amino acid sequence, and toxicity. Hepatotoxins mainly affect the liver, but it can affect other organs as well, being cytotoxic, dermatoxic, genotoxic, nephrotoxic, developmentally toxic, and potentially carcinogenic in extreme cases. Neurotoxins may lead to convulsion, tremor, paralysis, and death in minutes, infamously dubbed as a 'very fast death factor' (Vogiazi et al., 2019).

Detection of cyanotoxins can be done in a well-equipped certified laboratory using techniques, which are tedious, cumbersome, time and labor-intensive, and expensive. These techniques lack the required specificity and sensitivity. For example, animal models such as crustaceans, protozoans, insects, rotifers, cnidarians, nematodes, and oligochaetes were utilized for the detection of cyanotoxins. Cell-culture assays employing hepatocytes, intestine, fibroblasts, and neuroblastoma can be used as an alternative to animal-based assays. Biochemical assays include immunoassays (MCs/NOD, CYN, anatoxins, and STXs), enzymatic assays (enzyme-linked immunosorbent assays, ELISA), and receptor assays (PST, STX). Liquid chromatography (LC) coupled with photodiode array, mass spectrometry (MS), matrix-assisted laser desorption ionization/time-of-flight (MALDI-TOF), fluorescence, and electrochemistry are utilized for the quantification of (MCs, NOD, CYN, anatoxin). The fast-emerging environmental concern due to cyanotoxins requires the availability of affinity assays or sensing techniques that can detect the cyanotoxins onsite with high specificity, sensitivity in a real-time and cost-effective manner (Vogiazi et al., 2019).

A novel aptamer systematic evolution of ligand by exponential enrichment (SELEX) was devised using graphene oxide (GO) with high affinity and selectivity for CYN, and a combined fluorescent and electrochemical detection method with high binding affinities to ATX-a (Elshafey et al., 2014, 2015). Another aptamer utilizing a hapten-carrier conjugate as the SELEX target was employed for STX (Handy et al., 2013).

An impedance-based electrochemical biosensor was fabricated using a glassy carbon electrode (GCE) functionalized with AuNPs for the direct competitive detection of MC-LR. The biosensor exhibited very high sensitivity and specificity against MC-LR (Hou et al., 2016). A differential pulse voltammetry (DPV) biosensor employing immobilized graphene sheets on chitosan or glassy carbon electrodes (GCE) having conjugated carbon nanospheres with MAbs for signal amplification was deployed for the detection of MC-LR (He et al., 2017). An indirect competitive assay was developed using single-walled carbon nanohorns (SWNHs) functionalized on glassy carbon electrode (GCE) for the reliable and sensitive detection of MC-LR (Zhang et al., 2010). A DNAzyme based electrochemical immunosensor on carbon nanotubes (CNTs) was developed for the detection of MC-LR (Tian et al., 2014) (Fig. 4). A quantum dot (QD) mediated electrochemical immunosensor was fabricated for the indirect determination of MC-LR (Yu et al., 2009). A label-free immunosensor was designed using a glassy carbon electrode (GCE) composed of graphene-gold nanocomposites having carboxyl groups (polyDPB), gold nanoparticles layer (Au NP), an ionic liquid (IL) layer for the detection of MC-LR (Ruiyi et al., 2013). A biosensor based on an immune-field-effect transistor with a single-walled carbon nanotube (SWCNT) incorporating interdigitated electrodes (IDEs) was used for the MC-LR detection (Fu et al., 2016). Aptamer-based electrochemical biosensor was designed for the detection of multiple variants of MC-LR such as MC-LA, and MC-YR in a label-free manner (Ng et al., 2012). A label-free impedimetric aptasensor for ATX-a and STX was also reported (Handy et al., 2013; Elshafey et al., 2015).

4.8. Pathogens

Water-borne infections are the leading cause of mortality worldwide. These infections are spread by contaminated freshwater and wastewater sources. The pathogens common to both contaminated freshwater and wastewater are *Escherichia coli, Shigella dysenteriae, Salmonella typhi, Vibrio cholerae,*

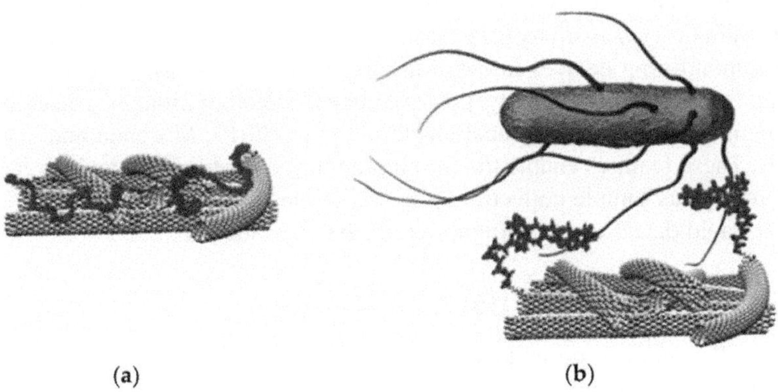

FIG. 4 Possible conformations of the aptamers that are self-assembled on carbon nanotubes (Zelada-Guillén et al., 2009; Mustafa et al., 2017). *Copyright © 2017 by the authors. Licensee MDPI, Basel, Switzerland. This article is an open-access article distributed under the terms and conditions of the Creative Commons Attribution (CC BY) license (http://creativecommons.org/licenses/by/4.0/).*

Campylobacter jejuni, Pseudomonas aeruginosa, and *Legionella* species. People from low- and middle-income groups are especially more vulnerable to water-borne infections owing to their exposure to insufficient and inadequate wastewater treatment and also to the leakage of water pipes that contaminate the water used for drinking and irrigation purposes. The untreated or partially treated wastewater is discharged into the water bodies such as rivers, ponds, etc. Domestic consumption of untreated or partially treated water for drinking, washing, and irrigation leads to a variety of water-borne infections to the exposed community (Rainbow et al., 2020).

Water-borne infections lead to approximately 600,000 deaths in the low- and middle-income populace annually and 1.5 million deaths globally as per the estimates of WHO (World Health Organization, 2014). This is especially more prevalent in Asia, Africa, and South America (Osiemo et al., 2019). Rapid urbanization and industrialization in limited space contribute to the larger effluent load from domestic and industrial settlements carrying a high nutrient load and microbial species from surface runoff. This surface runoff leads to the development of microbial communities causing disease. (Cabral, 2010; Pandey et al., 2014; Qin et al., 2014).

An early-warning device needs to be developed for the effective control and management of targeted contaminants. Such device should be capable of sensitive and selective detection of specific biomarkers, easy-to-use, resistant to biofouling, rapid response, end-user-type response, economical, and can be used on the site without the requirement of sophisticated instruments and trained and skilled manpower.

4.8.1 Lab-on-chip (LoC)

Electrochemical sensors offer portability, and miniaturization ability while maintaining high sensitivity. Electrochemical sensors can be used as hand-held, in situ monitoring devices integrating electrochemical sensing electrodes, microfluidics, signal connectivity, power sources, and other miniaturized elements for point-of-care (POC) applications (Jolly et al., 2019; Moschou and Tserepi, 2017). These features enable the development of an automated devices capable of autonomous sample collection, preparation, testing, analysis, and data readout for rapid detection (Domínguez et al., 2019; Mabey et al., 2004).

4.8.2 Lab-in-briefcase (LiB)

It is a self-contained portable system for on-site water monitoring. It contains the components required for water sample collection, sample pre-treatment, biosensing elements, and data processing units. The biosensing element is composed of a disposable biochip and reusable sensing device usually an electrochemical cell with a potentiostat. Single-use, low-cost, disposable electrochemical biochips are also available having microfabricated electrodes on screen-printed electrodes (SPEs); printed circuit boards (PCBs) as substrates, and evaporated metals (Ag, Au) on glass slides. However, evaporated Au, Ag electrodes on the glass surface are sensitive to physical abrasion and also lacks mechanical strength (Moschou and Tserepi, 2017; Smith et al., 2014).

4.8.3 Microfluidics

Integration of electrochemical sensors with microfluidics offers easy sample preparation for analyte detection. Lee et al. (2012) have devised a fully automated gradient-based platform having microfluidics embedded with a gold-patterened microarray chamber for the detection of multiple clinical biomarkers in an automated manner. This included gradient mixing of different solutions such as measurement buffers, antibody conjugated nanomaterials, and target antigens. Owen et al. (2013) have utilized a rotating magnetic disk and magnetic beads to mix complex analyte samples within a microfluidic device demonstrating sufficient sample mixing within a short span of time. Environment water samples are complex so require a necessary filtration step before proceeding for sample preparation and detection steps. Microfluidic on-chip devices including membrane or physical filtration, magnetic bead separation, and on-chip electrophoresis facilitate filtration of samples. Microfluidic electrophoretic methods such as gel electrophoresis, capillary electrophoresis, and dielectrophoresis are incorporated in microfluidics to enable the development of sensors with low sample requirements, are cost-effective, offer rapid detection, convenient integration, and sensing modules (Yu et al., 2016; Zhang et al., 2019).

One essential prerequisite for the on-site diagnosis is the automated preconcentration step or amplification of biological samples as these are present in low concentrations in natural samples like water. Lack of preconcentration of biological samples results in the decreased sensitivity of biosensors due to nonspecific binding (Contreras-Naranjo and Aguilar, 2019; Hook et al., 2014; Lichtenberg et al., 2019). Several protein preconcentration methods like solid-phase extraction (SPE), affinity-based magnetic bead enrichment, and isoelectric focusing (IEF) can be incorporated onto on-chip devices (Katoba et al., 2019; Sommer and Hatch, 2009; Whiteaker et al., 2007). For nucleic acid amplification, polymerase chain reaction (PCR), loop-mediated isothermal amplification (LAMP), and nucleic acid sequence-based amplification (NASBA) can be used (Fakruddin et al., 2013).

4.8.4 Electrochemical μPCR

It is an on-chip PCR miniaturization process including the integration of miniature heating elements and sensing elements with microfluidics (Ahrberg et al., 2016; Park et al., 2012; Wu et al., 2013a, b). Moschou et al. (2013) have developed a device with PCB as a substrate to integrate DNA amplification step with subsequent detection on Si-based array. Microfluidics was placed in a serpentine manner over resistance-based microheater elements of varying set temperatures. Samples were allowed to undergo 30 cycles of PCR. The amplified product is then sensed through a change in capacitance due to hybridization. Hwang et al. (2015) have used propylene and 400 μm double-sided tapes in place of microfluidic layers. These channels were placed over PCB-based μPCR with a heating circuit and thermal sensor. Electrophoretic results show similar amplification as obtained with conventional thermal-cycler PCR.

4.8.5 Bacteriophage

Carbon nanotubes (CNTs) with high electrical conductivity and large surface area are capturing the imagination of nanomaterial engineers worldwide for the design of electrochemical biosensors. In a study, single-walled CNTs (SWCNTs) were utilized to develop an immunosensor for the detection of bacteriophage MS2 from sewage impacted water supplies. Carboxy-functionalized SWCNTs were covalently attached to gold electrodes through carbodiimide coupling to cysteamine-modified gold electrodes. These SWCNTs were hydrazide functionalized by electrochemical grafting of diazonium salts. Subsequently, hydrazone bond formation led to the site-oriented immobilization of antibodies. Significant improvements in the signal-to-noise ratio, excellent sensitivity, and limit of detection (LOD) was achieved partly due to the ability of SWCNTs to transfer electrons with electroactive species and partly due to the large surface-to-volume ratio providing enough physical space for the immobilization of biomolecules. In another approach, SWCNTs were coupled with iron oxide nanoparticles functionalized with hydrazide groups for the immobilization of biomolecules (antibodies) through hydrazone coupling. This has facilitated MS2 isolation and concentration on an electrode surface. This has led to the minimization of nonspecific adsorption and matrix effects resulting in the further reduction of the limit of detection (LOD) through the incubation of magnetic immunocarriers with the large volumes of samples. Both approaches have successfully detected MS2 in sewage-contaminated water (Prieto-Simón et al., 2015).

4.8.6 Bacteria

A nanocomposite containing graphene oxide/carbon nanotube/poly (O-toluidine) (GO-CNT-POT) was designed for the reliable detection of Gram-positive bacteria *Bacillus subtilis*, Gram-negative bacteria *Escherichia coli,* and antibiotics Amoxicillin. The nanocomposite sensor was synthesized by in situ polymerization and characterized by X-ray powder diffractometry (XRD), and scanning electron microscopy (SEM). GO-CNT-POT nanocomposite was fabricated onto glassy carbon electrode (GCE) using conducting coating binders by the I-V technique. The GO-CNT-POT nanocomposite fabricated GCE electrodes show good selectivity, sensitivity, LOD, reproducibility, linear dynamic range, and electrochemical performances (Khan et al., 2016) (Fig. 4).

Singh et al. (2013) have utilized graphene oxide (GO)/chitosan (CHI) nanocomposite in an electrochemical genosensor for the detection of typhoid. The protocol involved covalent immobilization of *Salmonella typhi* 5′ amino-labeled ssDNA probe on GO-CHI/ITO via glutaraldehyde. Differential pulse voltammetry (DPV) gave good specificity and ability of ssDNA/GO-CHI/ITO to distinguish complementary, noncomplementary, and base mismatch sequence with detection in the range of 10 fM to 50 nM and LOD of 50 fM. This biosensor combined the advantage of large surface-to-volume ratio and good electrochemical activity of GO; good biocompatibility of chitosan enhancing DNA immobilization and electron transfer between DNA and electrode surface (ITO). Another assay developed by Wang et al. (2011) have utilized reduced graphene oxide (rGO) modified glassy-carbon electrode (GCE) to detect methicillin-resistant *Staphylococcus aureus* (MRSA) DNA employing electrochemical impedance spectroscopy (EIS) in a highly selective and sensitive manner. Jaiswal et al. (2018) have used carboxylated graphene nanoflakes (Cx-Gnfs) to fabricate an electrochemical genosensor for the detection of *Escherichia coli O157: H7* (*E. coli*). This platform has utilized Cx-Gnfs and r-GO as a sensing layer using electrochemical impedance spectroscopy (EIS). The sensor platform has retained 85% of its activity even after multiple regenerations (6-times) (Fig. 5).

Altintas et al. (2018) have fabricated a microfluidic-based electrochemical biosensor for the detection of *Escherichia coli* using nanomaterial-based immunoassay. The methodology was applied for the quantification of *E.coli*

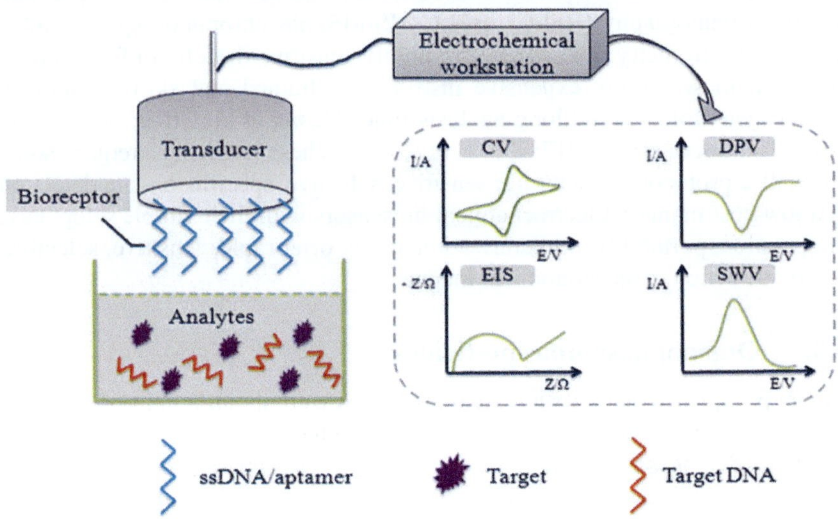

FIG. 5 Schematic diagram of electrochemical DNA biosensors (Wu et al., 2019). *Copyright © 2019 by the authors. Licensee MDPI, Basel, Switzerland. This article is an open-access article distributed under the terms and conditions of the Creative Commons Attribution (CC BY) license (http://creativecommons.org/licenses/by/4.0/).*

in the real water sample and it gave a similar detection limit. Cross-reactive studies by testing on *Shigella, Salmonella spp., Salmonella typhimurium,* and *Staphylococcus aureus* on *E. coli* specific antibody surface confirmed the high specificity of the immunoassay. The sensor surface was regenerated multiple times resulting in reduced cost of the assay.

4.9. Pesticides

Pesticides are most commonly used in the agriculture and food industry for the control, prevention, and elimination of pests. Pesticides can be classified as insecticide, herbicide, fungicide, bactericide depending upon the target pest species. Chemically, pesticides can be carbamates, dinitro compounds, organochlorines, organophosphates, pyrethroids, neonicotinoids, or triazines, etc. (World Health Organization, 2015). Clinical effects on humans can be in the form of neurotoxicity, carcinogenicity, and teratogenicity leading to cancer, birth defects, infertility, neurological disorders, allergy, and hypersensitivity. European Union allows the residual maximum limit (RML) to be 0.1 μg/L for individual pesticides while 0.5 μg/L for total pesticides (Oficial et al., 2013). Enzyme inhibition by the pesticide is the most common way of reliably detecting pesticide residues. For instance, organophosphorus and carbamates specifically inhibit the activity of acetylcholinesterase (AChE). According to an estimate of the United Nations, around 200,000 deaths occur annually on account of acute poisoning 99% of which occurs in developing nations. All this requires strict control in their release in the environment. Conventional techniques such as high-pressure liquid chromatography (HPLC), HPLC-MS/MS, gas chromatography coupled to mass spectrometry (GC-MS) offer highly sensitive detection of pesticides. However, these require expensive instruments, trained and skilled personal, high reagent volume, and long analysis time (Huang et al., 2019; Song et al., 2019; Timofeeva et al., 2017; Zhu et al., 2019). These limitations require some alternative protocols, which offer sensitive, selective, specific, on-site detection in a low-cost manner. Electrochemical biosensors with their simple setup, ease of operation, portability, miniaturization ability offer rapid, sensitive, selective, on-site detection in an economical manner.

4.9.1 Organophosporous pesticides

Organophosphates on oxidation yield highly toxic compounds dangerous to human health when enters human through food-chain. Electrochemical sensors are developed to biomonitor pesticide markers. Organophosphorus pesticides belong to the largest class of insecticides. These are involved in more poisoning cases than any other insecticides. These are potent inhibitors of the enzyme acetylcholinesterase (AChE), which destroys the neurotransmitter acetylcholine (Ach) in the nervous tissue. Therefore, the biomonitoring of organophosphorus insecticides requires the design of an electrochemical sensor that can monitor the activity of cholinesterase (ChE), acetylcholinesterase, or butyrylcholinesterase (Barry et al., 2009).

Diethylphosphorothionates constitute another major subclass of organophosphorus insecticides, such as chlorpyrifos, diazinon, and parathion. These undergo oxidative desulfation or diarylation to form chlorpyrifos-oxon (the neurotoxic moiety) or 3,5,6-trichloro-2-pyridinol (TCP) and diethyl thiophosphate respectively (Chambers and Chambers, 1989; Ma and Chambers, 1994). Therefore, the biomonitoring of Diethyl-phosphorothionates requires the development of an electrochemical sensor that can detect and quantify TCP (for references Barry et al., 2009). Exposure to the organophosphorus insecticides monitoring requires the monitoring of three biomarkers, viz. cholinesterase activity, chemical metabolites, and phosphorylated protein (Barry et al., 2009).

Cholinesterase activity can be determined by the electrochemical sensor having an enzyme-immobilized screen-printed carbon nanotube electrode (SPCNE). The sensing platform utilizes either cholinesterase with an appropriate substrate (Liu and Lin, 2005) or the cholinesterase combined with choline oxidase (ChO) to generate H_2O_2 as a byproduct for its detection (Guerrieri and Palmisano, 2001; Lin et al., 2004; Barry et al., 2009).

A competitive electrochemical immunoassay was developed using an enzyme-tagged analog for the direct detection and quantification of 3,5,6-trichloro-2-pyridinol (TCP), a metabolite marker for the insecticide chloropyrifos (Liu et al., 2005, 2006). Similarly, a quantum dot-tagged analog was utilized for the detection of Cd^{2+} ions for metabolite detection (Wu et al., 2007; Liu et al., 2007) (Barry et al., 2009). A disposable electrochemical enzymatic biosensor having AuNP functionalized screen-printed-electrode (AuNP-SPE) was utilized for the detection of paraoxon in pear and cucumber samples (Zhao et al., 2018).

Neonicotinoids such as Imidacloprid (IDP) are used to control agricultural pests like whiteflies, beetles, and lepidopterans. It enters the food chain through invertebrates and vertebrates and finally to humans in biomagnified amounts. An electrochemical sensor having reduced graphene oxide/cyclodextrin/glassy carbon electrode (*E*-rGO/CD/GCE) nanocomposite was designed for the detection of IDP in brown rice in a sensitive, selective, and reproducible manner. The electrochemical reduction approach used for the synthesis of electrodes is advantageous in terms of the economy of cost and time involved (Zhao et al., 2020).

4.9.2 Fungicide

Fungicides are the type of pesticides that prevent or control the growth of fungus and their spores. These are applied to control rust, smuts, blights, and mildews, etc. of plants. Fungicides work in a host-specific manner. However, mostly fungicides damage the fungal cell membrane or interfere with the energy production within the fungal cells. Broadly, fungicides serve three important purposes, viz., control and prevention of disease; increase crop productivity and reduce blemishes; and improve the storage life and quality of post-harvest produce. Mycotoxins are one of the most potent fungicides that has serious implications for both agriculture and public health. Notable examples of these mycotoxins are aflatoxins, ochratoxins, zearalenone, deoxynivalenol, fumonisins, citrinin, patulin, and ergot alkaloids. Aflatoxins and ochratoxins are potent carcinogens in humans. Additionally, ochratoxins lead to liver, kidney, and brain disorders in animals and humans. Zearalenone is an estrogen metabolite causing infertility in swine and poultry (Yin et al., 2008).

The advent of nanoparticle-based sensors has spearheaded the design and synthesis of sensors for soil and water sample analysis. These sensors are flexible, effective, and work with selective analytical methods, which are economical and are easy to operate. These sensors facilitate the analysis of trace amounts of pollutants from the environmental samples. These are easy to operate and less expensive. Lu and Gunasekaran (2019) have designed an electrochemical immunosensor for the simultaneous detection of two mycotoxins such as for fumonisin B1 (FB1) and deoxynivalenol (DON). The sensing platform included transparent indium-tin-oxide (ITO) coated glass integrated with capillary driven polydimethylsiloxane (PDMS) microfluidic channel with a detection limit in picogram/ml. A third-generation enzymatic biosensor was fabricated by Nieto et al. (2018) to quantify sterigmatocystin (STEH) involving GCE modified nanocomposite (GCE/SPE/rGO) of soybean peroxidase enzyme (SPE) and reduced graphene oxide (rGO). Similarly, a third-generation biosensor was designed to quantify STEH in corn samples spiked with mycotoxin utilizing the GCE modified nanocomposite of SPE and rGO (GCE/SPE/rGO). In a study by the Ilager group, a Tungsten oxide nanorod ($WO_3 \cdot 0.33H_2O$) based sensor was fabricated for the sensitive detection of fungicide, carbendazim (CBZ) with the lowest limit of detection (LOD) and limit of quantification (LOQ) (Ilager et al., 2021).

4.10. Soil microbes and pathogens

Soil microbes are crucial to various soil biological processes such as the degradation of organic matter, mineralization, and nutrient recycling. Thus, playing an important role in maintaining and sustaining the quality and functional diversity of the soil ecosystem. Enzymes released by these microbes are the quantitative indicator of the changes in the soil organic matter. One such marker enzyme is Beta-glucosidase (beta-Glu). Its activity provides a soil-quality indicator in terms of changes in the status of organic matter and its turnover. A simple assay using multi-walled carbon nanotubes (MWCNTs) was developed for the detection of glucose produced by the enzymatic degradation of cellobiose by beta-Glu. To achieve this, an assay platform was developed utilizing screen-printed carbon electrodes (SPCE) modified with multiwalled carbon nanotubes (SPCE-MWCNT) already containing co-immobilized glucose oxidase and horseradish peroxidase (HRP) enzymes. This has led to an improved amperometric response owing to the increased surface area upon electrode modification (Stege et al., 2010).

4.11. Nanotechnology for detection and remediation of environmental pollutants

Environment security is one of the fundamental requirements for the sustenance of planet Earth and life thereof. However, it remains one of the biggest challenges for humanity in the modern world. The post-industrialized world has seen the number of persistent and emerging chemicals polluting the environment and posing a serious threat to the ecology and ecosystem. Sensitive and rapid detection of these and their efficient removal from the environment is key for a clean, green, and sustainable environment. Nanomaterial-based electrochemical sensors are useful for the environmental monitoring and remediation of toxicants and pollutants (Zhang and Fang, 2010).

The advent of nanotechnology has given a fillip to the nanomaterial engineering and design of sensors for the real-time monitoring of fertilizers, herbicides, pesticides, insecticides, heavy metals, organic-inorganic pollutants, and pathogens (Chen and Yada, 2011; Kumar et al., 2015). These nanosensors can be fabricated for the intended use of monitoring physiochemical properties with greater sensitivity, selectivity, the lower limit of detection, portability, and on-site detection in a real-time manner (Fraceto et al., 2016; Liu et al., 2008). These nanosensors can be employed for the detection and monitoring of pathogens like bacteria, viruses, and fungus, etc. in addition to the monitoring of field conditions like moisture level, pH, temperature, nutrient status, and concentrations (Iavicoli et al., 2017).

4.12. Nanotechnology for water and soil remediation

One of the biggest sources of environmental pollution is water and soil contamination from natural or man-made sources. Hence, it is imperative to check these practices with the effective means of green technology to halt the uncontrolled spike in water and soil contamination. Nanotechnology has promising applications for providing sustainable solutions to the global problem of water and soil pollution and largely the environmental pollution. Nanosensors can remove heavy metals, metalloids, organic-inorganic pollutants, and dyes in addition to wastewater treatment in a cost-effective manner. Remediation through nanomaterials includes detoxification and transformation of pollutants. Nanomaterials because of their small size can penetrate and remain suspended on the subsurface for a longer duration as compared to the large-sized macroparticles, thus making the nano-remediation process more effective in terms of cost and time (Iavicoli et al., 2017). The nanomaterials that have been explored for this purpose are metal, metal-oxide and bimetallic NPs, carbon nanotubes, and nanofibers. Of these, nanoscale zero-valent iron (ZVI-NPs) owing to the low cost of their production and less toxic effects show great potential (Fu et al., 2014; Tosco et al., 2014; Yan et al., 2013). ZVI-NPs can degrade chlorinated compounds and heavy metals like Cd^{2+}, Cr^{6+} through redox reactions from tanneries and wastewaters; nitrogen from the activated sludge and phosphorous from the aqueous solutions (Lv et al., 2011; O'Carroll et al., 2013; Wu et al., 2013a, b; Liu et al., 2013). Metal oxide nanoparticles like Iron oxides, Magnetite acts as adsorbents owing to high affinity for the pollutants, redox cycle, magnetic properties, and ion exchange. After the removal of the contaminant the metal oxides can be recovered from the aqueous solutions (Braunschweig et al., 2013; Lee et al., 2013; Zhang et al., 2011, 2013). Nanostructured bimetallic systems—i.e., Pd/Fe, Ag/Fe, and Ni/Fe stabilized with carboxymethyl cellulose, polymers, and surfactants are employed for the removal of heavy metals, dyes, halogenated compounds and to neutralize bacteria. Porous Ti silicate and Al nanocomposite (Al_2O_3/TiO_2) can be used to remove heavy metals like Cd^{2+}, Pb^{2+}. However, the bimetallic nanoparticles are not overall safe for the field applications as they release reactive intermediates and toxicants as final products (Li et al., 2016; Singh et al., 2012; Zhou et al., 2011; Li et al., 2011; Yuan et al., 2012). Nanocomposites of the carbonaceous materials act as an adsorbent for halogenated compounds, polychlorinated biphenyls (PCB), polycyclic aromatic hydrocarbons, volatile organic compounds, herbicides, industrial dyes, and heavy metals in water (Farghali et al., 2013; Liu et al., 2011a, b; Hou et al., 2013; Hüffer et al., 2012; Moradi et al., 2011; Wang et al., 2014a; Wang et al., 2014b). Nanomaterials are useful for the treatment of environmental pollutants, but limitations are there in terms of regeneration, reusability, large-scale application, efficacy in wastewater, on-field applications, life-cycle, active working phase, and effect on different ecosystems, etc. There is an urgent need to thoroughly investigate their working mechanism, release effect, the fate of nanomaterials in the environment, recycling in the natural ecosystem, and effect on the human system through the food-chain. This is important and needs to be done as there are documented adverse effects of nanomaterials, especially those applied as nano-fertilizers, pesticides, and also for water and soil remediation (Iavicoli et al., 2011, 2012, 2013).

4.13. Nanoscale products for remediation

Cellulose nanomaterials provide a promising alternative as adsorbent for water remediation owing to large surface-area to volume-ratio, biocompatibility, and environmental inertness, etc. Cellulose nanomaterials provide a functionalized surfaces for the chemical moieties, which in turn increase the adsorption capacity of pollutants such as metal ions and organic contaminants (Hokkanen et al., 2016; Korhonen et al., 2011; Yu et al., 2013). A cellulose-coated fertilizer is used with controlled release and water-retention capacity serving as moisture-holding additive in the soil (Wu and Liu, 2008). However, as with other bionanomaterials, cellulosic nanomaterials such as the nanocrystalline and micro fibrillated cellulose have shown cytotoxicity, reduced cellular viability, and the health risk associated with the inhalation of nanocellulose (Clift et al., 2011; Pereira et al., 2013; Yang et al., 2013; Lin and Dufresne, 2014).

4.13.1 Nano-fertilizer

A nano fertilizer is released in a gradual, controlled, and targeted manner. This has an advantage over the conventional fertilizers or agrochemicals which do not reach the target effectively and hence are required to be given repeatedly that not only increases the cost but is also fraught with persistent side effects on the soil microbiota and humans. Nano fertilizers encapsulate more than one nutrient through the polymeric film of nanomaterials. This leads to the enhanced uptake of nutrients by the plants affecting the growth in a positive manner (De Rosa et al., 2010; Liu and Lal, 2015). Nanoporous zeolites impart high surface area and selectivity to the phyto-macronutrients causing its gradual release and subsequent uptake by the plant in a highly specific manner thus reducing the nutrient loss (Malekian et al., 2011; Zwingmann et al., 2011). Titanium dioxide (TiO_2) nanoparticles and carbon nanotubes (CNTs) enhance the germination and growth of plants due to light-induced generation of active oxygen like superoxide and hydroxide anions by TiO_2 nanoparticles.

4.13.2 Nano-pesticide

Nanopesticides offer the benefits of greater solubility, mobility, durability, bioavailability via nanoemulsions and nanodispersions (Suresh Kumar et al., 2013; Kah and Hofmann, 2014). Polymers of nanospheres and nanocapsules coupled with nanogels and nanofibers act as a reservoir of the active component for its slow and regulated release (Brunel et al., 2013). Similarly, other engineered nanomaterials like solid-lipid nanoparticles, liposomes, mesoporous silica have been studied as the carrier of pesticides (Kang et al., 2012; Ao et al., 2013). Silver nanoparticles (Ag-NP) are known inhibitors of plant pathogens when applied in a dose-dependent manner. Nano-formulations of silicon, TiO_2, and copper also perform the similar functions (Gogos et al., 2012; Iavicoli et al., 2017; Kah and Hofmann, 2014).

4.14. Risk assessment and management of nanomaterials

The risk assessment for the nanomaterials on human health is a tricky issue. This requires a detailed investigation of physical and chemical parameters, composition, morphology, functionalization, surface-area, particle size distribution, free and bound nanoparticles, bioavailability, bioactivity, toxicity, bioconcentration, bioaccumulation, sedimentation rate, and hazard profile to be assessed (Evans et al., 2013; Kookana et al., 2014). An additional threat is from "Biosolids," which contains nanosized metals usually from the consumer products (Ag, TiO_2, and ZnO-NPs). The implications of these nanomaterials, which are formed during the waste treatment process are not properly explored. These may further interact with the nanosized nutrients, and nanobiocidal substances posing a serious threat and risk to the ecosystem and human health. Biosolid derived nanomaterials may also enter the food web and may become the part of a plant, microbial or soil community getting biomagnified in the process and affecting the structure and function of the ecosystem. Initial toxicity testing has shown the pro-oxidant and pro-inflammatory effects of nanomaterials. It can cause adverse effects on the respiratory, cardiovascular, nervous systems, and endocrine systems (Iavicoli et al., 2013).

The risk assessment and management involve reliable, sensitive, selective, specific detection, quantification, and characterization of nanomaterials. It is important to find out those exposed and the level of exposure with their potential effects. The nano-formulations may have organic and inorganic constituents that undergo a transformation during synthesis, storage, or application due to physical, biological, and chemical processes occurring in the environment. Assays or techniques developed for their detection must be endowed with the ability to determine the changeable properties such as solubility, sorption, and degradation (Tiede et al., 2009). Current research is lacking in identifying the biomarkers for the engineered nanoparticles. Therefore, in vivo toxicokinetic studies are required to be done to identify possible biomarkers for validation and occupational biomonitoring field studies. Nanomaterials upon entering the body are translocated and distributed to different parts of the body. The fate of nanomaterials in vivo depends upon the inherent ability of the body to biotransform the nanomaterials for its subsequent elimination or interaction of the nanomaterials with the cellular materials for its upkeep (Kookana et al., 2014). Determining the exposure of the body to the nanomaterials and their inductive effect on the overall physiology especially the toxicity is required to be explored to build an adequate risk assessment process.

The management and control of nanomaterials-related risk require the development of a need-based strategy where exposure to the nanoparticles can be eliminated, and the application of state-of-art technology to monitor different stages of nanomaterials for its effective redressal may be established. Proof-of-concept from green chemistry should be applied for the design and synthesis of nanomaterials that do not add pollutants to the environment and pose health risk to the exposed populations. High-throughput screening should be done to investigate the pH of nanomaterial suspension, and oxidation state to eliminate probable interaction and subsequent integration of nanomaterials inside the body.

Regulatory measures through the appropriate legislations need to be put in place for providing the requisite permission for the design, extraction, and synthesis of nanomaterials and their use following due process of approval. The USA already exercises its control through EPA and FIFRA while Europe does this via "Plant protection and product regulation act." A joint effort from academia, industry, government and all stakeholders is needed to come up with a policy framework for its synthesis and usage (Iavicoli et al., 2017).

4.15. Conclusions and future perspectives

Increased environmental pollution has necessitated the development of a device that can identify the contaminants or pollutants in a simple, rapid, sensitive, selective, and specific manner without requiring expensive infrastructural setups and trained and skilled professionals. It should also facilitate on-site detection for effective and targeted control and management of pollutants. Electrochemical sensors exhibit all these attributes of a good sensing device and offer a wide range of applications for the determination of organic-inorganic pollutants from water sources, agricultural contaminants such as chemical fertilizers, pesticides, fungicides, etc. from the agricultural land. Electrochemical sensors can detect, and monitors water and soil-borne pathogens as well. Electrochemical sensors are minimally susceptible to the matrix effects and can detect the target analyte from among the background of a mixture of homologous and analogous analytes. However, electrochemical sensing devices need to be further optimized for their optimal performances for the broad range of biological matrices and their detection interferences from the biological matrices. The analytical performance of electrochemical sensors needs to be validated and cross-validated with other established and reliable analytical techniques such as mass spectrometry and ELISA. In vitro human and in vivo animal studies are required to be validated to assess the sensitivity, selectivity, specificity, and robustness of the biosensing platforms. Sensor performances need to be evaluated for the real-world samples in the ground-zero setups of the remote location for its universal applicability. Future beckons the electrochemical sensors with improvisations in the use of alternative affinity, sequestration stationary phases like titanium oxide, zirconium oxide, small nucleic acids strands (aptamers), chelators, self-assembled monolayers, polymer-modified nanoparticles, and nanostructured hydrogels. High-throughput efficiency could be achieved by developing multichannel, parallel, and multiplexed detection of multiple analytes.

References

Ahrberg, C., Manz, A., Chung, B., 2016. Polymerase chain reaction in microfluidic devices. Lab Chip 16, 3866–3884.

Altintas, Z., Akgun, M., Kokturk, G., Uludag, Y., 2018. A fully automated microfluidic-based electrochemical sensor for real-time bacteria detection. Biosens. Bioelectron. 100, 541–548.

Anmar, S., Oturan, N., Oturan, M.A., 2007. Electrochemical oxidation of 2-Nitrophenol in aqueous medium by electro-Fenton technology. J. Environ. Eng. Manag. 17, 89–96.

Ao, M., Zhu, Y., He, S., Li, D., Li, P., Li, J., Cao, Y., 2013. Preparation and characterization of 1-naphthylacetic acid-silica conjugated nanospheres for enhancement of controlled-release performance. Nanotechnology 24, 035601.

Ashley, K., 2003. Developments in electrochemical sensors for occupational and environmental health applications. J. Hazard. Mater. 102, 1–12.

Authier, L., Grossiord, C., Brossier, P., Limoges, B., 2001. Gold nanoparticle-based quantitative electrochemical detection of amplified human cytomegalovirus DNA using disposable microband electrodes. Anal. Chem. 73, 4450–4456.

Ayranci, R., Demirkol, D.O., Timur, S., Ak, M., 2017. Rhodamine-based conjugated polymers: potentiometric, colorimetric and voltammetric sensing of mercury ions in aqueous medium. Analyst 142 (18), 3407–3415.

Badihi-Mossberg, M., Buchner, V., Rishpon, J., 2007. Electrochemical biosensors for pollutants in the environment. Electroanalysis 19, 2015–2028.

Bard, A.J., Faulkner, L.R., 1980. Electrochemical Methods: Fundamentals and Applications. John Wiley and Sons Inc, New York, pp. 413–414.

Bard, A.J., Faulkner, L.R., 2001. Electrochemical Methods: Fundamentals and Applications. Wiley. 2001.

Barry, R.C., Lin, Y., Wang, J., Liu, G., Timchalk, C.A., 2009. Nanotechnology-based electrochemical sensors for biomonitoring chemical exposures. J. Expo. Sci. Environ. Epidemiol. 19 (1), 1–18.

Botasini, S., Heijo, G., Méndez, E., 2013. Toward decentralized analysis of mercury (II) in real samples. A critical review on nanotechnology-based methodologies. Anal. Chim. Acta 800, 1–11.

Braunschweig, J., Bosch, J., Meckenstock, R.U., 2013. Iron oxide nanoparticles in geomicrobiology: from biogeochemistry to bioremediation. New Biotechnol. 30, 793–802.

Brunel, F., El Gueddari, N.E., Moerschbacher, B.M., 2013. Complexation of copper(II) with chitosan nanogels: toward control of microbial growth. Carbohydr. Polym. 92, 1348–1356.

Cabral, J., 2010. Water microbiology. Bacterial pathogens and water. Int. J. Environ. Res. Public Health 7, 3657–3703.

Campàs, M., Garibo, D., Prieto-Simón, B., 2012. Novel nanobiotechnological concepts in electrochemical biosensors for the analysis of toxins. Analyst 137 (5), 1055–1067.

Chambers, J.E., Chambers, H.W., 1989. Oxidative desulfation of chlorpyrifos, chlorpyrifosmethyl, and leptophos by rat brain and liver. J. Biochem. Toxicol. 4 (1), 201–203.

Chen, H., Yada, R., 2011. Nanotechnologies in agriculture: new tools for sustainable development. Trends Food Sci. Technol. 22, 585–594.

Chen, S.H., Yuan, R., Chai, Y.Q., Zhang, L.Y., Wang, N., Li, X.L., 2007. Amperometric third-generation hydrogen peroxide biosensor based on the immobilization of hemoglobin on multiwall carbon nanotubes and gold colloidal nanoparticles. Biosens. Bioelectron. 22, 1268–1274.

Clever, G.H., Kaul, C., Carell, T., 2007. DNA–metal base pairs. Angew. Chem. Int. Ed. 46, 6226–6236.

Clift, M.J.D., Foster, E.J., Vanhecke, D., Studer, P., Wick, P., Gehr, P., Rothen-Rutishauser, B., Weder, C., 2011. Investigation of the interaction of cellulose nanofibers derived from cotton with a sophisticated human lung cell coculture. Biomacromolecules 12, 3666–3673.

Contreras-Naranjo, J., Aguilar, O., 2019. Suppressing non-specific binding of proteins onto electrode surfaces in the development of electrochemical immunosensors. Biosensors 9, 15.

Cui, R., Pan, H.C., Zhu, J.J., Chen, H.Y., 2007. Versatile immunosensor using CdTe quantum dots as electrochemical and fluorescent labels. Anal. Chem. 79, 8494–8501.

De Rosa, M.C., Monreal, C., Schnitzer, M., Walsh, R., Sultan, Y., 2010. Nanotechnology in fertilizers. Nat. Nanotechnol. 5, 91.

Domínguez, I., Oviedo-Ocaña, E., Hurtado, K., Barón, A., Hall, R., 2019. Assessing sustainability in rural water supply systems in developing countries using a novel tool based on multi-criteria analysis. Sustainability 11, 5363.

Elshafey, R., Siaj, M., Zourob, M., 2014. In vitro selection, characterization, and biosensing application of high-affinity cylindrospermopsin-targeting aptamers. Anal. Chem. 86 (18), 9196–9203.

Elshafey, R., Siaj, M., Zourob, M., 2015. DNA aptamers selection and characterization for development of label-free impedimetric aptasensor for neurotoxin Anatoxin-A. Biosens. Bioelectron. 68, 295–302.

Evans, D.E., Turkevich, L.A., Roettgers, C.T., Deye, G.J., Baron, P.A., 2013. Dustiness of fine and nanoscale powders. Ann. Occup. Hyg. 57, 261–277.

Fakruddin, M., Mannan, K., Hossain, M., Islam, S., Mazumdar, R., Chowdhury, A., Chowdhury, M., 2013. Nucleic acid amplification: alternative methods of polymerase chain reaction. J. Pharm. Bioallied Sci. 5, 245.

Farghali, A., Bahgat, M., ElRouby, W., Khedr, M., 2013. Decoration of multi-walled carbon nanotubes (MWCNTs) with different ferrite nanoparticles and its use as an adsorbent. J. Nanostruct. Chem. 3, 1–12.

Fraceto, L.F., Grillo, R., deMedeiros, G.A., Scognamiglio, V., Rea, G., Bartolucci, C., 2016. Nanotechnology in agriculture: which innovation potential does it have? Front. Environ. Sci. 4, 20.

Fu, F., Dionysiou, D.D., Liu, H., 2014. The use of zero-valent iron for groundwater remediation and wastewater treatment: a review. J. Hazard. Mater. 267, 194–205.

Fu, X., Zhao, C., Liu, X., Zhao, J., Niu, X., Zheng, L., Yang, Y., 2016. A layer-by-layer assembly label-free electrochemical immunosensor for the detection of microcystin-LR based on CHIT/PAMAM dendrimer/silver nanocubes. Int. J. Environ. Anal. Chem. 96 (3), 284–297.

Gogos, A., Knauer, K., Bucheli, T.D., 2012. Nanomaterials in plant protection and fertilization: current state, foreseen applications, and research priorities. J. Agric. Food Chem. 60, 9781–9792.

Guerrieri, A., Palmisano, F., 2001. An acetylcholinesterase/choline oxidase-based amperometric biosensor as a liquid chromatography detector for acetylcholine and choline determination in brain tissue. Anal. Chem. 73, 2875–2882.

Guo, S.J., Wang, E.K., 2007. Synthesis and electrochemical applications of gold nanoparticles. Anal. Chim. Acta 598, 181–192.

Han, C., Doepke, A., Cho, W., Likodimos, V., De La Cruz, A.A., Back, T., Heineman, W.R., Halsall, H.B., Shanov, V.N., Schulz, M.J.A., 2013. Multiwalled-carbon-nanotube-based biosensor for monitoring microcystin-LR in sources of drinking water supplies. Adv. Funct. Mater. 23 (14), 1807–1816.

Handy, S.M., Yakes, B.J., DeGrasse, J.A., Campbell, K., Elliott, C.T., Kanyuck, K.M., DeGrasse, S.L., 2013. First report of the use of a saxitoxin-protein conjugate to develop a DNA aptamer to a small molecule toxin. Toxicon 61 (1), 30–37.

Hanrahan, G., Patil, D.G., Wang, J., 2004. Electrochemical sensors for environmental monitoring: design, development and applications. J. Environ. Monit. 6, 657–664.

He, Z., Wei, J., Gan, C., Liu, W., Liu, Y., 2017. A rolling circle amplification signal-enhanced immunosensor for ultrasensitive microcystin-LR detection based on a magnetic GrapheneFunctionalized electrode. RSC Adv. 7 (63), 39906–39913.

Hokkanen, S., Bhatnagar, A., Sillanpää, M., 2016. A review on modification methods to cellulose-based adsorbents to improve adsorption capacity. Water Res. 91, 156–173.

Hook, S., Gallagher, E., Batley, G., 2014. The role of biomarkers in the assessment of aquatic ecosystem health. Integr. Environ. Assess. Manag. 10, 327–341.

Hou, L., Ding, Y., Zhang, L., Guo, Y., Li, M., Chen, Z., Wu, X., 2016. An ultrasensitive competitive immunosensor for impedimetric detection of microcystin-LR via antibody-conjugated enzymatic biocatalytic precipitation. Sensors Actuators B Chem. 233, 63–70.

Hou, L., Zhu, D., Wang, X., Wang, L., Zhang, C., Chen, W., 2013. Adsorption of phenanthrene, 2-naphthol, and 1-naphthylamine to colloidal oxidized multiwalled carbon nanotubes: effects of humic acid and surfactant modification. Environ. Toxicol. Chem. 32, 493–500.

Huang, Y., Shi, T., Luo, X., Xiong, H., Min, F., Chen, Y., 2019. Determination of multi-pesticide residues in green tea with a modified QuEChERS protocol coupled to HPLC-MS/MS. Food Chem. 275, 255–264.

Hüffer, T., Kah, M., Hofmann, T., Schmidt, T.C., 2012. How redox conditions and irradiation affect sorption of PAHs by dispersed fullerenes (NC60). Environ. Sci. Technol. 47, 6935–6942.

Hussain, M.M., Rahman, M.M., Asiri, A.M., 2016. Efficient 2-nitrophenol chemical sensor development based on Ce2O3 nanoparticles decorated CNT nanocomposites for environmental safety. PLoS One. https://doi.org/10.1371/journal.pone.0166265.

Hussain, M.M., Rahman, M.M., Asiri, A.M., 2017. Ultrasensitive and selective 4-aminophenol chemical sensor development based on nickel oxide nanoparticles decorated carbon nanotube nanocomposites for green environment. J. Environ. Sci. (China) 53, 27–38.

Hwang, J., Kim, S., Kim, Y., Song, H., Park, C., Kim, J., 2015. Implementation of PCB-based PCR chip using double-sided tape. Int. J. Control Autom. 8, 117–124.

Iavicoli, I., Fontana, L., Leso, V., Bergamaschi, A., 2013. The effects of nanomaterials as endocrine disruptors. Int. J. Mol. Sci. 14, 16732–16801.

Iavicoli, I., Leso, V., Bergamaschi, A., 2012. Toxicological effects of titanium dioxide nanoparticles: a review of *in vivo* studies. J. Nanomater., 964381. 36pp.

Iavicoli, I., Leso, V., Fontana, L., Bergamaschi, A., 2011. Toxicological effects of titanium dioxide nanoparticles: a review of in vitro mammalian studies. Eur. Rev. Med. Pharmacol. Sci. 15, 481–508.

Iavicoli, I., Veruscka, L., Beezholdb, D.H., Shvedova, A.A., 2017. Nanotechnology in agriculture: opportunities, toxicological implications, and occupational risks. Toxicol. Appl. Pharmacol. 329, 96–111.

Ilager, D., Seo, H., Kalanur, S.S., Shetti, N.P., Aminabhavi, T.M., 2021. A novel sensor based on $WO_3{\cdot}0.33H_2O$ nanorods modified electrode for the detection and degradation of herbicide, carbendazim. J. Environ. Manag. 279, 111611.

Isichei, T.O., Okieimen, F.E., 2014. Adsorption of 2-nitrophenol onto water hyacinth activated carbon-kinetics and equilibrium studies. Environ. Pollut. 3.

Jaiswal, N., Pandey, C., Soni, A., Tiwari, I., Rosillo-Lopez, M., Salzmann, C., Malhotra, B., Sumana, G., 2018. Electrochemical genosensor based on carboxylated graphene for detection of water-borne pathogen. Sensors Actuators B Chem. 275, 312–321.

Ji, R., Niu, W., Chen, S., Xu, S., Ji, X., Yuan, L., Zhao, H., Geng, M., Qiu, J., Li, C., 2019. Target-inspired Pb^{2+}-dependent DNAzyme for ultrasensitive electrochemical sensor based on MoS_2-AuPt nanocomposites and hemin/G-quadruplex DNAzyme as signal amplifier. Biosens. Bioelectron. 144, 111560.

Jolly, P., Rainbow, J., Regoutz, A., Estrela, P., Moschou, D., 2019. A PNA-based Lab-on-PCB diagnostic platform for rapid and high sensitivity DNA quantification. Biosens. Bioelectron. 123, 244–250.

Kah, M., Hofmann, T., 2014. Nanopesticide research: current trends and future priorities. Environ. Int. 63, 224–235.

Kang, M.A., Seo, M.J., Hwang, I.C., Jang, C., Park, H.J., Yu, Y.M., Youn, Y.N., 2012. Insecticidal activity and feeding behavior of the green peach aphid, Myzus persicae, after treatment with nano types of pyrifluquinazon. J. Asia Pac. Entomol. 15, 533–541.

Katoba, J., Kuupiel, D., Mashamba-Thompson, T.P., 2019. Toward improving accessibility of point-of-care diagnostic services for maternal and child health in low-and middle-income countries. Point Care 18, 17–25.

Khan, A.A.P., Khan, A., Rahman, M.M., Asiri, A.M., Oves, M., 2016. Lead sensors development and antimicrobial activities based on graphene oxide/carbon nanotube/poly(O-toluidine) nanocomposite. Int. J. Biol. Macromol. 89, 198–205.

Kookana, R.S., Boxall, A.B., Reeves, P.T., Ashauer, R., Beulke, S., Chaudhry, Q., Cornelis, G., Fernandes, T.F., Gan, J., Kah, M., Lynch, I., Ranville, J., Sinclair, C., Spurgeon, D., Tiede, K., Van den Brink, P.J., 2014. Nanopesticides: guiding principles for regulatory evaluation of environmental risks. J. Agric. Food Chem. 62, 4227–4240.

Korhonen, J.T., Kettunen, M., Ras, R.H.A., Ikkala, O., 2011. Hydrophobic nanocellulose aerogels as floating, sustainable, reusable, and recyclable oil absorbents. ACS Appl. Mater. Interfaces 3, 1813–1816.

Kumar, P., Kim, K.H., Deep, A., 2015. Recent advancements in sensing techniques based on functional materials for organophosphate pesticides. Biosens. Bioelectron. 70, 469–481.

Lee, M., Lee, K., Kim, K., Oh, K., Choo, J., 2012. SERS-based immunoassay using a gold array-embedded gradient microfluidic chip. Lab Chip 12, 3720.

Lee, H.U., Lee, S.C., Lee, Y.C., Vrtnik, S., Kim, C., Lee, S., Lee, Y.B., Nam, B., Lee, J.W., Park, S.Y., Lee, S.M., Lee, J., 2013. Sea-urchin-like iron oxide nanostructures for water treatment. J. Hazard. Mater. 262, 130–136.

Li, Q., Chen, X., Zhuang, J., Chen, X., 2016. Decontaminating soil organic pollutants with manufactured nanoparticles. Environ. Sci. Pollut. Res. Int. 23, 11533–11548.

Li, T., Dong, S., Wang, E., 2010. A lead(II)-driven DNA molecular device for turn-on fluorescence detection of lead(II) ion with high selectivity and sensitivity. J. Am. Chem. Soc. 132, 13156–13157.

Li, Z., Yuan, S., Wan, J., Long, H., Tong, M., 2011. A combination of electrokinetics and Pd/Fe PRB for the remediation of pentachlorophenol-contaminated soil. J. Contam. Hydrol. 124, 99–107.

Lichtenberg, J., Ling, Y., Kim, S., 2019. Non-specific adsorption reduction methods in biosensing. Sensors 19, 2488.

Lin, N., Dufresne, A., 2014. Nanocellulose in biomedicine: current status and future prospect. Eur. Polym. J. 59, 302–325.

Lin, Y.Y., Liu, G.D., Wai, C.M., Lin, Y.H., 2007. Magnetic beads-based bioelectrochemical immunoassay of polycyclic aromatic hydrocarbons. Electrochem. Commun. 9, 1547–1552.

Lin, Y.H., Lu, F., Wang, J., 2004. Disposable carbon nanotube modified screen-printed biosensor for amperometric detection of organophosphorus pesticides and nerve agents. Electroanalysis 16, 145–149.

Lin, Y.H., Yantasee, W., Wang, J., 2005. Carbon nanotubes CNTs for the development of electrochemical biosensors. Front. Biosci. 10, 492–505.

Liu, H., Chen, H., Zou, X., Xie, Q., Qing, C., Chen, D., Frost, R.L., 2013. Removal of phosphorus using NZVI derived from reducing natural goethite. Chem. Eng. J. 234, 80–87.

Liu, F., Chung, S., Oh, G., Seo, T.S., 2011a. Three-dimensional graphene oxide nanostructure for fast and efficient water-soluble dye removal. ACS Appl. Mater. Interfaces 4, 922–927.

Liu, R., Lal, R., 2015. Potentials of engineered nanoparticles as fertilizers for increasing agronomic productions. Sci. Total Environ. 514, 131–139.

Liu, G.D., Lin, Y.H., 2005. Electrochemical sensor for organophosphate pesticides and nerve agents using zirconia nanoparticles as selective sorbents. Anal. Chem. 77, 5894–5901.

Liu, G., Lin, Y.Y., Wang, J., Wu, H., Chien, M.W., Lin, Y., 2007. Disposable electrochemical immunosensor diagnosis device based nanoparticle probe and immunochromatographic strip. Anal. Chem. 79, 7644–7653.

Liu, G.D., Riechers, S.L., Timchalk, C., Lin, Y., 2005. Sequential injection/electrochemical immunoassay for quantifying the pesticide metabolite 3,5,6-trichloro-2-pyridinol. Electrochem. Commun. 7, 1463–1470.

Liu, Y., Su, G., Zhang, B., Jiang, G., Yan, B., 2011b. Nanoparticle-based strategies for detection and remediation of environmental pollutants. Analyst 136, 872–877.

Liu, G., Timchalk, C., Lin, Y., 2006. Bioelectrochemical magnetic immunosensing of trichloropyridinol: a potential insecticide biomarker. Electroanalysis 18 (16), 1605–1613.

Liu, Y., Tong, Z., Prud'homme, R.K., 2008. Stabilized polymeric nanoparticles for controlled and efficient release of bifenthrin. Pest Manag. Sci. 64, 808–812.

Long, F., Zhu, A., Shi, H., Wang, H., Liu, J., 2013. Rapid on-site/in-situ detection of heavy metal ions in environmental water using a structure-switching DNA optical biosensor. Sci. Rep. 3, 2308.

Loyprasert, S., Thavarungkul, P., Asawatreratanakul, P., Wongkittisuksa, B., Limsakul, C., Kanatharana, P., 2008. Label-free capacitive immunosensor for microcystin-LR using self-assembled thiourea monolayer incorporated with ag nanoparticles on gold electrode. Biosens. Bioelectron. 24 (1), 78–86.

Lu, L., Gunasekaran, S., 2019. Dual-channel ITO-microfluidic electrochemical immunosensor for simultaneous detection of two mycotoxins. Talanta 194, 709–716.

Lv, X., Xu, J., Jiang, G., Xu, X., 2011. Removal of chromium(VI) from wastewater by nanoscale zero-valent iron particles supported on multiwalled carbon nanotubes. Chemosphere 85, 1204–1209.

Ma, T., Chambers, J.E., 1994. Kinetic parameters of desulfuration and dearylation of parathion and chlorpyrifos by rat liver microsomes. Food Chem. Toxicol. 32 (8), 763–767.

Mabey, D., Peeling, R., Ustianowski, A., Perkins, M., 2004. Diagnostics for the developing world. Nat. Rev. Microbiol. 2, 231–240.

Malekian, R., Abedi-Koupai, J., Eslamian, S.S., 2011. Influences of clinoptilolite and surfactant-modified clinoptilolite zeolite on nitrate leaching and plant growth. J. Hazard. Mater. 185, 970–976.

Mishra, G.K., Barfidokht, A., Tehrani, F., Mishra, R.K., 2018. Food safety analysis using electrochemical biosensors. Foods 7, 141.

Mohammed, M., Rahman, M.M., Hussain, M.M., Asiri, A.M., 2017. Fabrication of 3-methoxyphenol sensor based on Fe3O4 decorated carbon nanotube nanocomposites for environmental safety: real sample analyses. PLoS One. https://doi.org/10.1371/journal.pone.0177817.

Moradi, O., Yari, M., Moaveni, P., Norouzi, M., 2011. Removal of p-nitrophenol and naphthalene from petrochemical wastewater using SWCNTs and SWCNT-COOH surfaces. Fullerenes Nanotubes Carbon Nanostruct. 20, 85–98.

Moschou, D., Tserepi, A., 2017. The lab-on-PCB approach: tackling the μTAS commercial upscaling bottleneck. Lab Chip 17, 1388–1405.

Moschou, D., Vourdas, N., Filippidou, M., Tsouti, V., Kokkoris, G., Tsekenis, G., Zergioti, I., Chatzandroulis, S., Tserepi, A., 2013. Integrated biochip for PCR-based DNA amplification and detection on capacitive biosensors. In: SPIE Proc. vol. 8765.

Mustafa, F., Hassan, R.Y.A., Andreescu, S., 2017. Multifunctional nanotechnology-enabled sensors for rapid capture and detection of pathogens. Sensors 17, 2121.

Ng, A., Chinnappan, R., Eissa, S., Liu, H., Tlili, C., Zourob, M., 2012. Selection, characterization, and biosensing application of high affinity congener-specific microcystin-targeting aptamers. Environ. Sci. Technol. 46 (19), 10697–10703.

Nieto, C.D., Granero, A.M., Garcia, D., Nesci, A., Barros, G., Zon, M.A., Fernandez, H., 2018. Development of a third-generation biosensor to determine sterigmatocystin mycotoxin: an early warning system to detect aflatoxin B1. Talanta 194, 253–258.

Niu, P., Gich, M., Roig, A., Fernández-Sánchez, C., 2018. Metal nanoparticle carbon gel composites in environmental water sensing applications. Chem. Rec. 18 (7–8), 749–758.

O'Carroll, D., Sleep, B., Krol, M., Boparai, H., Kocur, C., 2013. Nanoscale zero valent iron and bimetallic particles for contaminated site remediation. Adv. Water Resour. 51, 104–122.

Oficial, D., Comisi, D.E.L.A., Europeo, P., Europeo, P., Alimentarius, C., 2013. Scientific support for preparing an EU position for the 45th session of the codex committee on pesticide residues (CCPR). EFSA J. 11, 1–5.

Ono, A., Cao, S., Togashi, H., Tashiro, M., Machinami, T., Oda, S., Miyake, Y., Okamoto, I., Tanaka, Y., 2008. Specific interactions between silver(i) ions and cytosine-cytosine pairs in DNA duplexes. Chem. Commun. 44, 4825–4827.

Osiemo, M., Ogendi, G., M'Erimba, C., 2019. Microbial quality of drinking water and prevalence of water-related diseases in Marigat Urban Centre, Kenya. Environ. Health Insights 13, 117863021983698.

Owen, D., Mao, W., Alexeev, A., Cannon, J., Hesketh, P., 2013. Microbeads for sampling and mixing in a complex sample. Micromachines 4, 103–115.

Pandey, P., Kass, P., Soupir, M., Biswas, S., Singh, V., 2014. Contamination of water resources by pathogenic bacteria. AMB Express 4, 51.

Park, C., Kim, J., Ku, J., Kim, Y., Song, H., Kim, J., 2012. Printed circuit board-based polymerase chain reaction chip. Sens. Lett. 10, 1197–1202.

Patolsky, F., Zheng, G.F., Lieber, C.M., 2006. Fabrication of silicon nanowire devices for ultrasensitive, labelfree, real-time detection of biological and chemical species. Nat. Protoc. 1, 1711–1724.

Pereira, M.M., Raposo, N.R., Brayner, R., Teixeira, E.M., Oliveira, V., Quintao, C.C., Camargo, S., Mattoso, L.H., Brandao, H.M., 2013. Cytotoxicity and expression of genes involved in the cellular stress response and apoptosis in mammalian fibroblast exposed to cotton cellulose nanofibers. Nanotechnology 24, 075103.

Pokhrel, L.R., Ettore, N., Jacobs, Z.L., Zarr, A., Weir, M.H., Scheuerman, P.R., Kanel, S.R., Dubey, B., 2017. Novel carbon nanotube (CNT)-based ultrasensitive sensors for trace mercury(II) detection in water: a review. Sci. Total Environ. 574, 1379–1388.

Prieto-Simón, B., Bandaru, N.M., Saint, C., Voelcker, N.H., 2015. Tailored carbon nanotube immunosensors for the detection of microbial contamination. Biosens. Bioelectron. 67, 642–648.

Qin, H., Su, Q., Khu, S., Tang, N., 2014. Water quality changes during rapid urbanization in the Shenzhen Rivercatchment: an integrated view of socio-economic and infrastructure development. Sustainability 6, 7433–7451.

Rahman, M.R., Balkhoyor, H.B., Asiri, A.M., 2017. Phenolic sensor development based on chromium oxide-decorated carbon nanotubes for environmental safety. J. Environ. Manag. 188, 228–237.

Rahman, M.M., Jamal, A., Khan, S.B., Faisal, M., 2011. Fabrication of highly sensitive ethanol chemical sensor based on Sm-doped Co3O4 nanokernels by a hydrothermal method. J. Phys. Chem. C 115, 9503.

Rahman, M.M., Khan, A., Asiri, A.M., 2015. Chemical sensor development based on poly (O anisidine)silverized MWCNT nanocomposites deposited on glassy carbon electrodes for environmental remediation. RSC Adv. 5, 71370.

Rainbow, J., Sedlackova, E., Jiang, S., Maxted, G., Moschou, D., Richtera, L., Estrela, P., 2020. Integrated electrochemical biosensors for detection of waterborne pathogens in low-resource settings. Biosensors 10, 36.

Rebeiro, S., Silva, A.M.T., Figueiredo, J.L., Faria, J.L., 2013. Removal of 2-nitrophenol by catalytic wet peroxide oxidation using carbon materials with different morphological and chemical properties. Appl. Catal. B Environ. 140–141, 356–362.

Renedo, O.D., Alonso-Lomillo, M.A., Martinez, M.J.A., 2007. Recent developments in the field of screen-printed electrodes and their related applications. Talanta 73, 202–219.

Reverté, L., Prieto-Simón, B., Campàs, M., 2016. New advances in electrochemical biosensors for the detection of toxins: nanomaterials, magnetic beads and microfluidics systems. A Review. Anal. Chim. Acta 908, 8–21.

Ruiyi, L., Qianfang, X., Zaijun, L., Xiulan, S., Junkang, L., 2013. Electrochemical immunosensor for ultrasensitive detection of microcystin-LR based on graphene-gold nanocomposite/functional conducting polymer/gold nanoparticle/ionic liquid composite film with electrodeposition. Biosens. Bioelectron. 44 (1), 235–240.

Sadik, O.A., Van Emon, J.M., 1996. Applications of electrochemical immunosensors to environmental monitoring. Biosens. Bioelectron. 11 (8), 1–9.

Singh, R., Misra, V., Mudiam, M.K.R., Chauhan, L.K.S., Singh, R.P., 2012. Degradation of -HCH spiked soil using stabilized Pd/FeO bimetallic nanoparticles: pathways, kinetics and effect of reaction conditions. J. Hazard. Mater. 237–238, 355–364.

Singh, A., Sinsinbar, G., Choudhary, M., Kumar, V., Pasricha, R., Verma, H., Singh, S., Arora, K., 2013. Graphene oxide-chitosan nanocomposite based electrochemical DNA biosensor for detection of typhoid. Sensors Actuators B Chem. 185, 675–684.

Smith, J.P., Foster, C.W., Metters, J.P., Sutcliffe, O.B., Banks, C.E., 2014. Metallic impurities in graphene screen-printed electrodes can influence their electrochemical properties. Electroanalysis 26, 2429–2433.

Sole, S., Alegret, S., 2001. Environmental toxicity monitoring using electrochemical biosensing systems. Environ. Sci. Pollut. Res. Int. 8, 256–264.

Somerset, V.S., Hernandez, L.H., Iwuoha, E.I., 2011. Stripping voltammetric measurement of trace metal ions using screen-printed carbon and modified carbon paste electrodes on river water from the Eerste-Kuils River system. J. Environ. Sci. Health A Tox. Hazard. Subst. Environ. Eng. 46 (1), 17–32.

Sommer, G.J., Hatch, A.V., 2009. IEF in microfluidic devices. Electrophoresis 30, 742–757.

Song, N., Lee, J.Y., Mansur, A.R., Jang, H.W., Lim, M., Lee, Y., 2019. Determination of 60 pesticides in hen eggs using the QuEChERS procedure followed by LC-MS/MS and GC-MS/MS. Food Chem. 298, 125050.

Stege, P.W., Messina, G.A., Bianchi, G., Olsina, R.A., Raba, J., 2010. Determination of beta-glucosidase activity in soils with a bioanalytical sensor modified with multiwalled carbon nanotubes. Anal. Bioanal. Chem. 397 (3), 1347–1353.

Suresh Kumar, R.S., Shiny, P.J., Anjali, C.H., Jerobin, J., Goshen, K.M., Magdassi, S., Mukherjee, A., Chandrasekaran, N., 2013. Distinctive effects of nano-sized permethrin in the environment. Environ. Sci. Pollut. Res. Int. 20, 2593–2602.

Tang, C.X., Zhao, Y., He, X.W., Yin, X.B., 2010. A "turn-on" electrochemiluminescent biosensor for detecting Hg21 at femtomole level based on the intercalation of Ru(phen)321 into ds-DNA. Chem. Commun. 46, 9022–9024.

Tian, J., Zhao, H., Yuan, F., Quan, X., Chen, S., 2014. Ultrasensitive immunoassay of microcystins-LR using G-Quadruplex DNAzyme as an electrocatalyst. Int. J. Environ. Anal. Chem. 94 (10), 988–1000.

Tiede, K., Hassellov, M., Breitbarth, E., Chaudhry, Q., Boxall, A.B.A., 2009. Considerations for environmental fate and ecotoxicity testing to support environmental risk assessments for engineered nanoparticles. J. Chromatogr. A 1216, 503–509.

Timofeeva, I., Shishov, A., Kanashina, D., Dzema, D., Bulatov, A., 2017. On-line in-syringe sugaring-out liquid-liquid extraction coupled with HPLC-MS/MS for the determination of pesticides in fruit and berry juices. Talanta 167, 761–767.

Tosco, T., Papini, M.P., Viggi, C.C., Sethi, R., 2014. Nanoscale zerovalent iron particles for groundwater remediation: a review. J. Clean. Prod. 77, 10–21.

US EPA, 2001. Mercury Update: Impact of Fish Advisories. EPA Fact Sheet EPA-823-F-01-011. EPA, Office of Water, Washington, DC.

Vogiazi, V., Cruz, A., Mishra, S., Shanov, V., Heineman, W.R., Dionysiou, D.D., 2019. A comprehensive review: development of electrochemical biosensors for detection of cyanotoxins in freshwater. ACS Sens. 4 (5), 1151–1173.

Wang, J., Chen, Z., Chen, B., 2014b. Adsorption of polycyclic aromatic hydrocarbons by graphene and graphene oxide nanosheets. Environ. Sci. Technol. 48, 4817–4825.

Wang, F., Haftka, J.J.H., Sinnige, T.L., Hermens, J.L.M., Chen, W., 2014a. Adsorption of polar, nonpolar, and substituted aromatics to colloidal graphene oxide nanoparticles. Environ. Pollut. 186, 226–233.

Wang, J., Li, R., 1990. On-valve electrochemical detector for high-speed flow injection analysis. Anal. Chem. 62, 2414–2416.

Wang, J., Liu, G.D., Lin, Y.H., 2006. Amperometric choline biosensor fabricated through electrostatic assembly of bienzyme/polyelectrolyte hybrid layers on carbon nanotubes. Analyst 131, 477–483.

Wang, Z., Zhang, J., Chen, P., Zhou, X., Yang, Y., Wu, S., Niu, L., Han, Y., Wang, L., Chen, P., Zhou, X., Yang, Y., Wu, S., Niu, L., Han, Y., Wang, L., Chen, P., Boey, F., Zhang, Q., Liedberg, B., Zhang, H., 2011. Label-free, electrochemical detection of methicillin-resistant staphylococcus aureus DNA with reduced graphene oxide-modified electrodes. Biosens. Bioelectron. 26, 3881–3886.

Whiteaker, J.R., Zhao, L., Zhang, H.Y., Feng, L.C., Piening, B.D., Anderson, L., Paulovich, A.G., 2007. Antibody-based enrichment of peptides on magnetic beads for mass-spectrometry-based quantification of serum biomarkers. Anal. Biochem. 362, 44–54.

Willner, I., Zayats, M., 2007. Electronic aptamer-based sensors. Angew. Chem. Int. Ed. 46, 6408–6418.

World Health Organization, 2014. UNICEF. Progress on Drinking Water and Sanitation. World Health Organization, Geneva, Switzerland.

World Health Organization, 2015. Human Biomonitoring: Facts and Figures. WHO Regional Office for Europe, Copenhagen, Denmark.

Wu, J., Fu, Z.F., Yan, F., Ju, H.X., 2007. Biomedical and clinical applications of immunoassays and immunosensors for tumor markers. Trac-Trends Anal. Chem. 26, 679–688.

Wu, J., Kodzius, R., Cao, W., Wen, W., 2013a. Extraction, amplification and detection of DNA in microfluidic chip-based assays. Microchim. Acta 181, 1611–1631.

Wu, L., Liu, M., 2008. Preparation and characterization of cellulose acetate-coated compound fertilizer with controlled-release and water-retention. Polym. Adv. Technol. 19, 785–792.

Wu, D., Shen, Y., Ding, A., Mahmood, Q., Liu, S., Tu, Q., 2013b. Effects of nanoscale zerovalent iron particles on biological nitrogen and phosphorus removal and microorganisms in activated sludge. J. Hazard. Mater. 262, 649–655.

Wu, Q., Zhang, Y., Yang, Q., Yuan, N., Zhang, W., 2019. Review of electrochemical DNA biosensors for detecting food borne pathogens. Sensors (Basel) 19 (22), 4916.

Xiao, Y., Rowe, A.A., Plaxco, K.W., 2007. Electrochemical detection of parts-perbillion lead via an electrode-bound DNAzyme assembly. J. Am. Chem. Soc. 129, 262–263.

Xu-Cheng, F., Wu, J., Nie, L., Cheng-Gen, X., Jin-Huai, L., Xing-Jiu, H., 2012. Electropolymerized surface ion imprinting films on a gold nanoparticles/single-wall carbon nanotube nanohybrids modified glassy carbon electrode for electrochemical detection of trace mercury(II) in water. Anal. Chim. Acta 720, 29–37.

Yan, W., Lien, H.L., Koel, B.E., Zhang, W.X., 2013. Iron nanoparticles for environmental clean-up: recent developments and future outlook. Environ. Sci. 15, 63–77.

Yang, X., Bakaic, E., Hoare, T., Cranston, E.D., 2013. Injectable polysaccharide hydrogels reinforced with cellulose nanocrystals: morphology, rheology, degradation, and cytotoxicity. Biomacromolecules 14, 4447–4455.

Yin, Y.N., Yan, L.Y., Jiang, J.H., Ma, Z.H., 2008. Biological control of aflatoxin contamination of crops. J Zhejiang Univ Sci B 9 (10), 787–792.

Yu, H.W., Lee, J., Kim, S., Nguyen, G.H., Kim, I.S., 2009. Electrochemical immunoassay using quantum dot/antibody probe for identification of cyanobacterial hepatotoxin microcystin-LR. Anal. Bioanal. Chem. 394 (8), 2173–2181.

Yu, X., Tong, S., Ge, M., Wu, L., Zuo, J., Cao, C., Song, W., 2013. Adsorption of heavy metal ions from aqueous solution by carboxylated cellulose nanocrystals. J. Environ. Sci. 25, 933–943.

Yu, S., Xu, J., Huang, K., Chen, J., Duan, J., Xu, Y., Qing, H., Geng, L., Deng, Y., 2016. A novel method to predict protein aggregations using two-dimensional native protein microfluidic chip electrophoresis. Anal. Methods 8, 8306–8313.

Yuan, S., Long, H., Xie, W., Liao, P., Tong, M., 2012. Electrokinetic transport of CMC-stabilized Pd/Fe nanoparticles for the remediation of PCP-contaminated soil. Geoderma 185–186, 18–25.

Zacco, E., Galve, R., Marco, M.P., Alegret, S., Pividori, M.I., 2007. Electrochemical biosensing of pesticide residues based on affinity biocomposite platforms. Biosens. Bioelectron. 22, 1707–1715.

Zaib, M., Athar, M.M., Saeed, A., Farooq, U., 2015. Electrochemical determination of inorganic mercury and arsenic—a review. Biosens. Bioelectron. 74, 895–908.

Zelada-Guillén, G.A., Riu, J., Düzgün, A., Rius, F.X., 2009. Immediate detection of living bacteria at ultralow concentrations using a carbon nanotube based potentiometric aptasensor. Angew. Chem. Int. Ed. 48, 7334–7337.

Zhang, H., Chang, H., Neuzil, P., 2019. DEP-on-a-Chip: dielectrophoresis applied to microfluidic platforms. Micromachines 10, 423.

Zhang, L., Fang, M., 2010. Nanomaterials in pollution trace detection and environmental improvement. Nano Today 5, 128–142.

Zhang, J., Lei, J., Xu, C., Ding, L., Ju, H., 2010. Carbon nanohorn sensitized electrochemical immunosensor for rapid detection of microcystin-LR. Anal. Chem. 82 (3), 1117–1122.

Zhang, J., Zhai, S., Li, S., Xiao, Z., Song, Y., An, Q., Tian, G., 2013. Pb(II) removal of Fe3O4@ SiO2–NH2 core–shell nanomaterials prepared via a controllable sol–gel process. Chem. Eng. J. 215–216, 461–471.

Zhang, F., Zhu, Z., Dong, Z., Cui, Z., Wang, H., Hu, W., Zhao, P., Wang, P., Wei, S., Li, R., Ma, J., 2011. Magnetically recoverable facile nanomaterials: synthesis, charac-terization and application in remediation of heavy metals. Microchem. J. 98, 328–333.

Zhao, F., Yao, Y., Li, X., Lan, L., Jiang, C., Ping, J., 2018. Metallic transition metal dichalcogenide nanosheets as an effective and biocompatible transducer for electrochemical detection of pesticide. Anal. Chem. 90, 11658–11664.

Zhao, Y., Zheng, X., Wang, Q., Zhe, T., Bai, Y., Bu, T., Zhang, M., Wang, L., 2020. Electrochemical behavior of reduced graphene oxide/cyclodextrins sensors for ultrasensitive detection of imidacloprid in brown rice. Food Chem. 333, 127495.

Zhou, H., Han, J., Baig, S.A., Xu, X., 2011. Dechlorination of 2,4-dichlorophenoxyaceticacid by sodium carboxymethyl cellulose-stabilized Pd/Fe nanoparticles. J. Hazard. Mater. 198, 7–12.

Zhu, Z., Tao, L., Li, F., 2014. 2-Nitrophenol reduction promoted by S. putrefaciens 200 and biogenic ferrous iron: the role of different size-fractions of dissolved organic matter. J. Hazard. Mater. 279, 436–443.

Zhu, B., Xu, X., Luo, J., Jin, S., Chen, W., Liu, Z., 2019. Simultaneous determination of 131 pesticides in tea by on-line GPC-GC—MS/MS using graphitized multi-walled carbon nanotubes as dispersive solid phase extraction sorbent. Food Chem. 276, 202–208.

Zwingmann, N., Mackinnon, I.D.R., Gilkes, R.J., 2011. Use of a zeolite synthesised from alkali treated kaolin as a K fertiliser: glasshouse experiments on leaching and uptake of K by wheat plants in sandy soil. Appl. Clay Sci. 53, 684–690.

Chapter 5

Electrochemical biosensors: Biomonitoring of food adulterants, allergens, and pathogens

5.1. Introduction

Food waste is a chronic and persistent problem that is plaguing the human population globally. The annual loss of food waste per year is around 1.3 billion tons costing 1 trillion dollars in terms of energy and water resources. This also translates into food-related global health burden especially in developing and under-developed nations (World Health Organization (WHO), 2015; Castell-Perez et al., 2017). Persistent increase in the human population has breached the carrying capacity of the planet Earth and has put tremendous strain on the food production and supply to sustain the ever-increasing population. Problems are manifold arising out of sufficient food production for all, minimizing food waste on account of pest infestation, packaging, and food processing industries. Among all these, the contamination of food by pathogens and adulteration of food with spurious sub-standard substances is another challenge that needs to be addressed at the war-footing. A coordinated and cooperative effort is required from the stakeholders to address the problems of food safety and security, effective and efficient management of food distribution, packaging and supply, environment and agricultural sustainability to ensure uninterrupted food supply, minimizing food waste, generation of novel foodstuffs, etc. (Neethirajan et al., 2018a, b).

Seamless production of food supply depends on a multitude of factors such as electricity, irrigation, fertile land, and sufficient requisite infrastructure. A robust financial system is necessary for the supply of funds and resource mobilization for its sustenance. Institutionalized support in the form of favorable legislation on the part of governance is necessary for the growth of agriculture and hence the food supply. Technological innovation in terms of improvised agricultural equipment, techniques, hybrid seeds, and technological innovations for the post-agricultural harvest is key to ensure the food supply chain (Neethirajan et al., 2018a, b). Concerted efforts are already under operation to

Electrochemical Biosensors. https://doi.org/10.1016/B978-0-323-90632-6.00001-7

identify the drivers of food production. Data mining or knowledge discovery in databases (KDD) is key in elucidating the inter-relationship between food production, food safety and welfare, environmental issues, and climate change (Garcia, 2013).

Food safety is the prime concern for human and animal health. Conventional techniques in food safety analysis are tedious, cumbersome, labor- and time-intensive, and cost-ineffective. Additionally, it requires trained and skilled man-power and sophisticated infrastructural setup. This chapter emphasizes the role electrochemical biosensors play in the detection and biomonitoring of food adulterants, food allergens, foodborne pathogens, food toxins, and quality control in the food and beverage industry ensuring food safety and monitoring of potential toxic effects of nanomaterials utilized for biosensor fabrication. It also focuses on other roles, such as advanced food packaging processes, managing wastage of food, and nanomaterial-driven biosensing in food engineering. The development and evolution of electrochemical biosensing platforms incorporating nanomaterials for the biomonitoring of all aspects of the food industry are analyzed. Strengths and limitations of the electrochemical sensing platforms and their future are discussed in the food industry including food manufacturing, processing, packaging, and transportation down to consumption. This chapter aims to serve as a complete reference material and an important source of information to all stakeholders including scientists, engineers, and farmers, etc. for augmenting their knowledge on the existing state of electrochemical biosensing for the detection and monitoring of food quality, adulterants, allergens, toxins, pathogens, packaging, and waste management, etc.

The key focus of the chapter is on the following (i) electrochemical biosensors, their major types, (ii) use of different nanomaterials for the advancement of electrochemical sensors, (iii) biosensors for the biomonitoring of food adulterants, (iv) biosensors for the biomonitoring of food allergens, (vi) biosensors for the biomonitoring of food-borne pathogens, (vii) biosensors for the biomonitoring of food-toxins, (viii) Quality control in food and beverage industry, (ix) biomonitoring of food safety, (x) nanomaterial toxicity, (xi) food packaging, (xii) food waste management, (xiii) biosensing in food engineering, and (xiv) future directions.

5.1.1 Biosensors

A biosensor is an analytical device equipped with a biorecognition element (aptamer, antibody, enzyme, etc.) for quantifying the target molecule of interest. Signal (physiochemical or biological) generated by the interaction between recognition element and target analyte, quantified by the transducer is measured. Signals can be optical (colorimetric, fluorescence, chemiluminescence, and surface plasmon resonance) and electrical (voltammetry, impedance, and capacitance). Based on the type of transducer element, biosensors can be of three types: electrical, optical, and mechanical. Biosensors can also be classified into electrochemical biosensors, piezoelectric biosensors, calorimetric biosensors, and optical biosensors based on transducers (Wu et al., 2019a, b). Optical transducers for the fabrication of biosensors is internal reflection, fluorescence resonance energy transfer (FRET), chemiluminescence, bioluminescence, surface plasmon resonance (SPR), localized SPR (LSPR), imaging SPR (iSPR), and resonance-enhanced absorption (Jayan et al., 2019). The biosensor can be of various types depending upon the choice of bioreceptors such as antibody biosensors, DNA biosensors, enzyme biosensors, whole-cell biosensors, and phage biosensors (Fig. 1).

5.1.2 Electrochemical biosensors

Electrochemical biosensors can be of two basic types, viz. biocatalytic and affinity sensors. Biocatalytic sensors analyze the reaction between the enzyme and the target to produce electroactive molecules while the affinity sensors monitor the reaction between the bioreceptor and target molecule to quantify a signal. Aptasensors, immunosensors, and genosensors fall under the category of affinity biosensors (Santos et al., 2012). Electrochemical biosensors can be labeled or label-free depending upon the use of labels to enhance the signal. Electrochemical biosensors show distinct advantages over optical biosensors in terms of analytical accuracy especially of complex food matrices, portability, miniaturization, rapidity, and cost involved. Food allergens and adulterants are detected by voltammetry, impedimetric, and amperometry-based

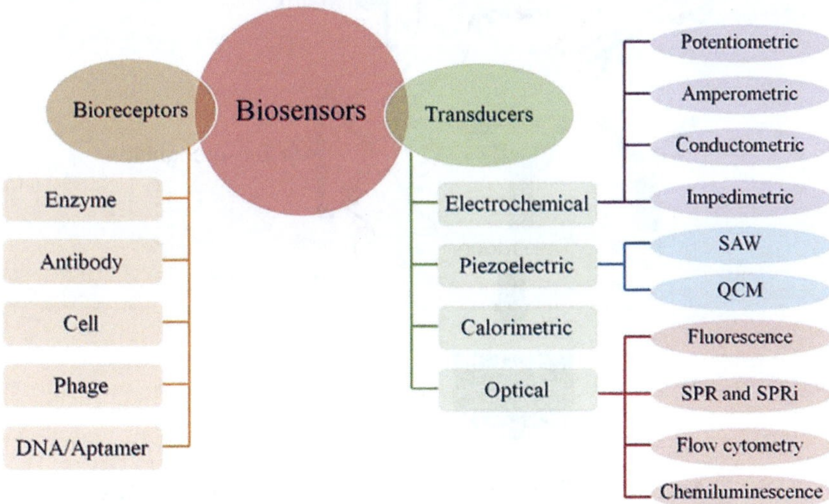

FIG. 1 Components and classification of the biosensor (Wu et al., 2019a, b). *Copyright © 2019 by the authors. Licensee MDPI, Basel, Switzerland. This article is an open-access article distributed under the terms and conditions of the Creative Commons Attribution (CC BY) license (http://creativecommons.org/licenses/by/4.0/).*

electrochemical techniques (Neethirajan et al., 2018a, b). Mechanical biosensors measure the mass-sensitive sensor surface deflection as target analytes is immobilized onto the sensor surface (Haddada et al., 2018). Assays based on mechanical sensors are affinity assays, fingerprint assays, separation assays, and spectrometric assays (Arlett et al., 2011). Quartz crystal microbalance (QCM) is a mechanical biosensor. It can detect the shift in mass in sub-nanogram quantity. The change in mass is recognized by the resonant frequency of quartz crystal utilized for the sensitive detection and quantification of whole-cell microorganisms (Yu et al., 2018) (Fig. 2).

Recent development in the field of electrochemical biosensing involves augmentation of specificity, sensitivity, and response time. The selectivity of the transducer can be enhanced by electrochemical modifications that involve electrochemical pre-treatment for enhancing the electrochemical properties of electrode surface (Hayat et al., 2014). Aptamers offer higher sensitivity and selectivity as compared to antibodies, scale-up production in an economical in vitro system, and enhanced environmental stability. Aptamers can be easily integrated into an electrochemical biosensing platform. Aptamers with optical transducers require high precision, expensive instrumentation, and sophisticated numerical algorithms to interpret the data. Aptamers coupled with electrochemical transducers generate electric signal read-out to provide a simple,

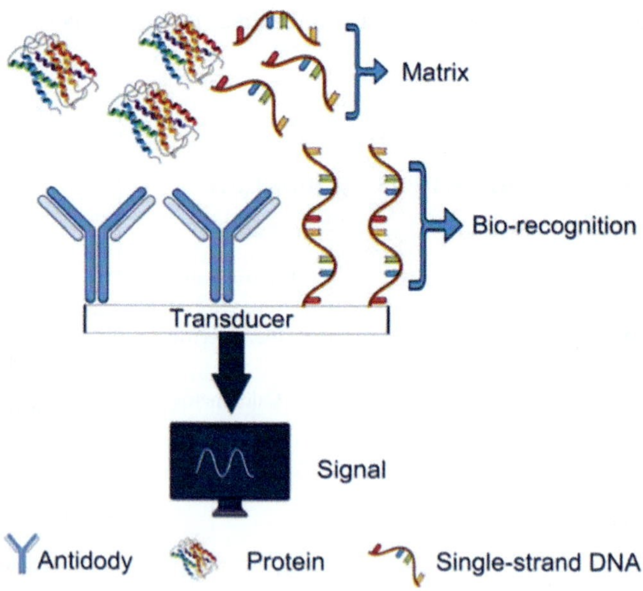

FIG. 2 Schematic diagram of the components of a biosensor (Adriano Aquino and Conte-Junior, 2020). *Copyright © 2020 by the authors. Licensee MDPI, Basel, Switzerland. This article is an open-access article distributed under the terms and conditions of the Creative Commons Attribution (CC BY) license (http://creativecommons.org/licenses/by/4.0/).*

economical, and accurate sensing platform for diverse applications (Rapini and Marrazza, 2017).

5.1.3 Nanomaterial aided advancement in electrochemical biosensors

The ever-expanding domain of nanotechnology and nanomaterial engineering is finding promising applications in real-world problems like health, energy, environment, and agriculture, etc. Nanotechnology-based interventions in agriculture in the field of food production, crop protection, pathogen and toxin detection, water purification, food packaging, wastewater treatment, and environmental remediation have enhanced productivity and efficiency.

Nanomaterials find important applications in the field of analytical chemistry as it improves the sensitivity, selectivity, specificity, and limit of detection down to femtomolar concentrations. Nanomaterial-based biosensors are advantageous over conventional biosensing techniques such as electrochemical, fluorescence, ultra-violet (UV)-Vis, and high-performance liquid chromatography (HPLC) owing to rapid, specific, sensitive, and selective detection in the real-time and economical manner (Alarcon-Angeles et al., 2016). Single and array nanochannels made up of graphene are used to improve biosensor performance. Single nanochannels are employed for drilling nanopores on silicon oxide/nitride membranes using transmission/scanning electron microscopy (TEM/SEM) techniques and electron-beam lithography, and for insertion into lipid-bilayer to mimic pore-forming toxins. Nanochannel arrays, such as micro- or nano-molding high-ordered mesoporous thin films, metallic substrate anodization, and anodized aluminum oxide nanoporous membranes are increasingly used owing to their advantage of mass production, easy functionalization, and impedimetric, voltammetric, capacitive, conductometric, and resistive measurements (Neethirajan et al., 2018a, b) (Fig. 3).

Au-based nanomaterials are advantageous in terms of optical, electrical, catalytical properties, and biocompatibility. Au nanomaterials find interesting applications in multimodal nanocomposites preparation, enhance the signal of plasmonic colorimetric analysis to develop multifunctional Au-nanomaterials, and nanochannels (Wang et al., 2017). Inorganic nanomaterials and their derivatives are used in various biosensing applications. However, their safety, security, and toxic properties are the cause of concern. Surface plasmon resonance (SPR) and quartz crystal microbalance (QCM) based affinity sensing are utilized for the analysis of food quality in a label-free and real-time manner (Scarano et al., 2016).

Recently, different nanomaterials such as zero-dimensional (0D) nanoparticles (NPs, including nanodots), 1D nanorods (containing nanowires and nanotubes), 2D nanosheets, and 3D metal-organic frameworks (MOFs) have been successfully utilized to meet the challenges and requirements of advanced nano-biosensing methods (Yang et al., 2016). 2D nanomaterial having a large surface-to-volume ratio coupled with ultrathin planar nanostructures provides numerous sites for the attachment of analytes. 2D nanomaterial exhibits higher conductivity due to tunable electronic configuration and resulting bandgap variation (Xia et al., 2014). 2D nanomaterials show good compatibility with silicon channel technology facilitating device miniaturization (Tyagi et al., 2020). 2D nanomaterials exhibit good mechanical strengths and optical properties. However, the desired physiochemical properties can be enhanced by the introduction of defect engineering, doping, and fine-tuning of structural properties during preparation (Liu et al., 2016a, b). Examples of 2D nanomaterials include graphene family nanomaterials, MXene, transition metal dichalcogenides, single-elemental layered crystalline materials, and metal oxides. 2D nanomaterials in conjunction with nanocomposites on integration with transducers can enhance the analytical performance of the sensor (for references Raja et al., 2021) (Figs. 3 and 4).

5.2. Biomonitoring of food adulterants

Food fraud constitutes a serious health problem globally of which food adulteration comprises 3% of the total food-fraud events. It occurs when there is

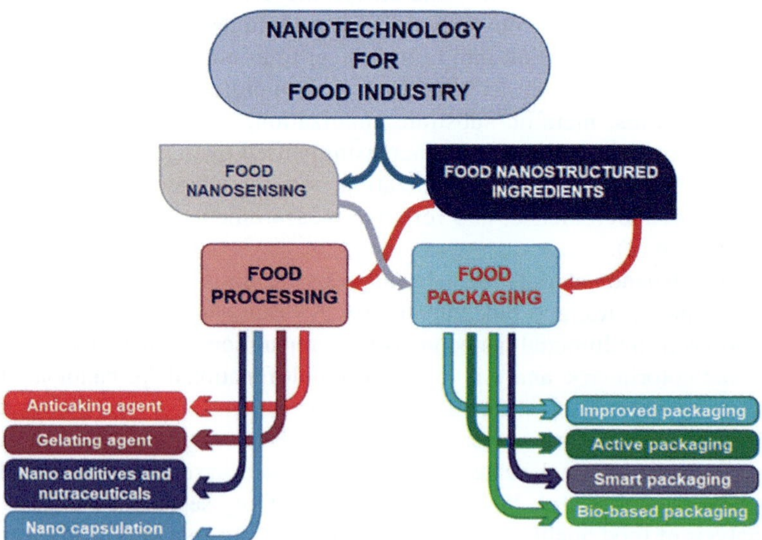

FIG. 3 Application of (bio)nanotechnology in different fields of the food industry (Primož et al., 2021). *Copyright © 2021 by the authors. Licensee MDPI, Basel, Switzerland. This article is an open-access article distributed under the terms and conditions of the Creative Commons Attribution (CC BY) license (http://creativecommons.org/licenses/by/4.0/).*

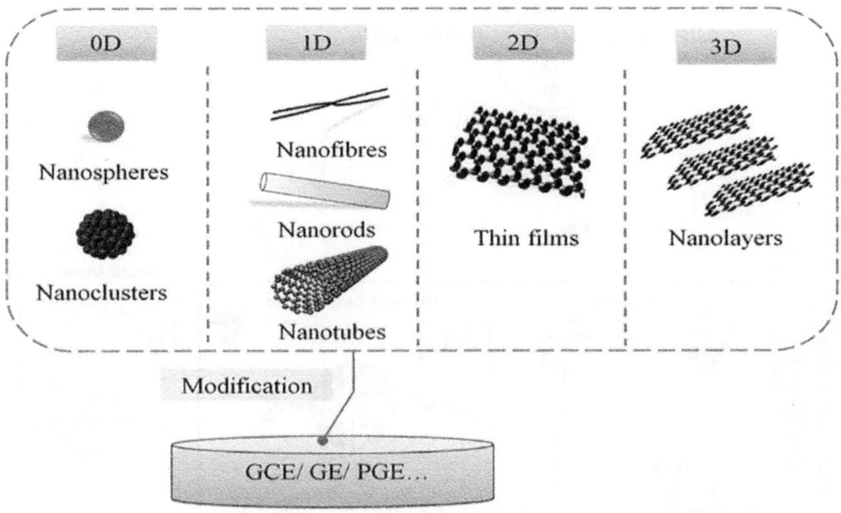

FIG. 4 Nanomaterials are commonly used in the modification of electrochemical DNA biosensors (Wu et al., 2019a, b). *Copyright © 2019 by the authors. Licensee MDPI, Basel, Switzerland. This article is an open-access article distributed under the terms and conditions of the Creative Commons Attribution (CC BY) license (http://creativecommons.org/licenses/by/4.0/).*

deliberate and intentional tampering in terms of mixing, addition, and substitution to garner economic benefit.

An assay involving DNA/RNA hybridization on the surface of strep-MBs was developed for the detection of horse meat in beef meat. The resulting heteroduplex was labeled with selective antibody and bacterial antibody binding protein to detect the Fc region of the mammalian IgG. The assay reliably detected the horse meat by recognizing the D-loop region of the horse mtDNA (Valat et al., 2000; Akram et al., 2004) (Fig. 5).

A switch-based electrochemical sensor having a self-assembled thiolated redox (methylene blue)-tagged DNA probe onto gold electrodes was developed to detect melamine in milk. DNA triplex, which is formed in the presence of melamine kept the MB away from the electrode generating measurable lower square wave voltammetry (SWV) oxidation response (for references Campuzano et al., 2020).

The ethanol content of alcoholic beverages can be determined by electrochemical sensors using an immobilized alcohol oxidase enzyme. For example, an alcohol dehydrogenase sensor was developed on a multiwalled carbon nanotube (MWCNT) (Santos et al., 2006). Bisphenol A (BPA) is used for the manufacture of polycarbonate and epoxy resin, which is further utilized for the production of water bottles, feeding bottles, coating materials, and food cans. An enzyme-based nanoparticle functionalized biosensor with either Ni-nanoparticles or Fe_2O_3 was used for the detection of BPA (Deng et al., 2014).

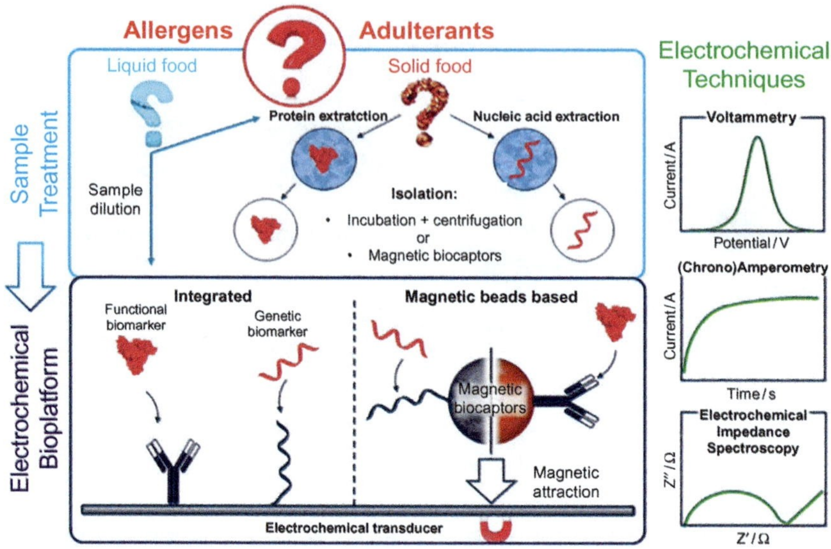

FIG. 5 Schematic display of the fundamentals (target biomarker extraction and electrochemical detection) involved in the construction of electrochemical biosensors for the determination of food allergens and adulterants at a genetic or functional level using integrated or magnetic beads (MBs)-based strategies (Campuzano et al., 2020).

5.2.1 Glutamate

Monosodium glutamate is a popular food additive. Glutamate is harmful to humans as it causes muscle atrophy, neurotoxicity, and damage to the brain, liver, and kidney. It contaminates water on secretion from the human body and inhibits plant growth. Glutamate is present in variable concentrations in different food sources, hence difficult to detect and quantify. Current biosensing platforms identify glutamate as glutamate dehydrogenase (GLDH) or glutamate oxidase (GluOx) as a bio-recognition elements. Biosensors involving GluOx as biorecognition element are simpler than those involving GLDH due to the requirement of additional NAD+ as a cofactor. Current biosensing platforms for glutamate detection are plagued by the high potential requirements for NADH oxidation on the electrode, high cost of enzymes, and lower stability (Hughes et al., 2016).

5.2.2 Genetically modified organisms (GMO)

Genetically modified organisms (GMO) are the much-touted product of a biotechnological revolution in agriculture during the 1990s. Currently, more than 50% cotton, 45% soybean, and 40% corn are the product of genetic modifications. However, GMOs are not good for human consumption in entirety due to their adverse effect on the gastrointestinal tract, antibiotic resistance, allergy, and the undesired gene flow to other organisms. Biosensors detect GMOs in food products through the identification of GM genes employing isothermal DNA amplification, and rapid signal detection, etc. (Meriç et al., 2014).

Detection of GM food utilizes PCR that analyzes specific sequences of DNA. However, increasing demand for portability has led to instrument miniaturization for its on-site use. DNA chips are now used for the on-site analysis of samples. SERS barcoded nanosensors with encapsulated AuNPs and silica, conjugated to oligonucleotides were used for the detection of *Bacillus thuringiensis (Bt)* gene transformed rice expressing insecticidal proteins (Chen et al., 2012).

5.3. Biomonitoring of food allergens

Allergy is a hypersensitive immune response in response to allergen-specific IgE. It can be type I (immediate hypersensitivity) or type II (delayed hypersensitivity) depending upon the development of symptoms on exposure to the allergens. Food-borne allergy is a common global affliction, prevalent more in the industrialized developed world. It affects children (6%–8%) more than the adults (4%). The most common allergens accounting for almost 90% of allergies worldwide are eggs, milk, crustaceans, shellfish, fish, peanuts, nuts, soybeans, and wheat. Therefore, food labeling showing relevant information about allergens is crucial to the sensitive customers (Campuzano et al., 2020).

Allergic consumers face the additional threat of undeclared allergic substances through adulteration, inappropriate practices, and cross-contamination. Cross-contamination may occur during the manufacturing process itself due to insufficient cleaning, or during the transfer or storage process. This can be mitigated by having different production lines for different products. However, storage of different products at the same place poses the possibility of cross-contamination (Eissa et al., 2017).

Allergens can be cross-reactive, requiring multiplexed detection strategies that can identify multiple biomarkers simultaneously. Determination of allergens can be achieved through immunological methods or DNA-based assays. It can be direct where allergens are detected or indirect where a biomarker diagnostic of an allergen is detected. Immunological methods like ELISA depend on the binding between the target molecule and immunoglobulin. These protocols are time- and labor-intensive requiring trained manpower and sophisticated laboratory setup. DNA-based assays employ RT-PCR and have the advantage of detecting multiple analytes. However, it is irrelevant for the food that contains more protein and lower DNA content such as eggs (Alves et al., 2016).

Optical and electrochemical biosensors owing to their low cost of detection, simple framework, portability with other analytical devices, and automation can play important role in identifying food-borne allergens and prophylaxis. Electrochemical biosensors show a distinct advantage over optical biosensors owing to their low cost, portability, and analytical precision for the complex food matrices. Voltammetry, impedimetric, and amperometry are favorable electrochemical techniques for allergen detection (Bunney et al., 2017; Eissa et al., 2017; Lin et al., 2017; Malvano et al., 2019) (Fig. 6). Magnetic particles coupled with screen-printed electrodes (SPEs) can be molded to different geometry, miniaturized, and multiplexed to work with very small sample volumes. This reduces the matrix effect, assay-time, and allows the detection in a high-throughput manner (Kudr et al., 2018; Pastucha et al., 2019).

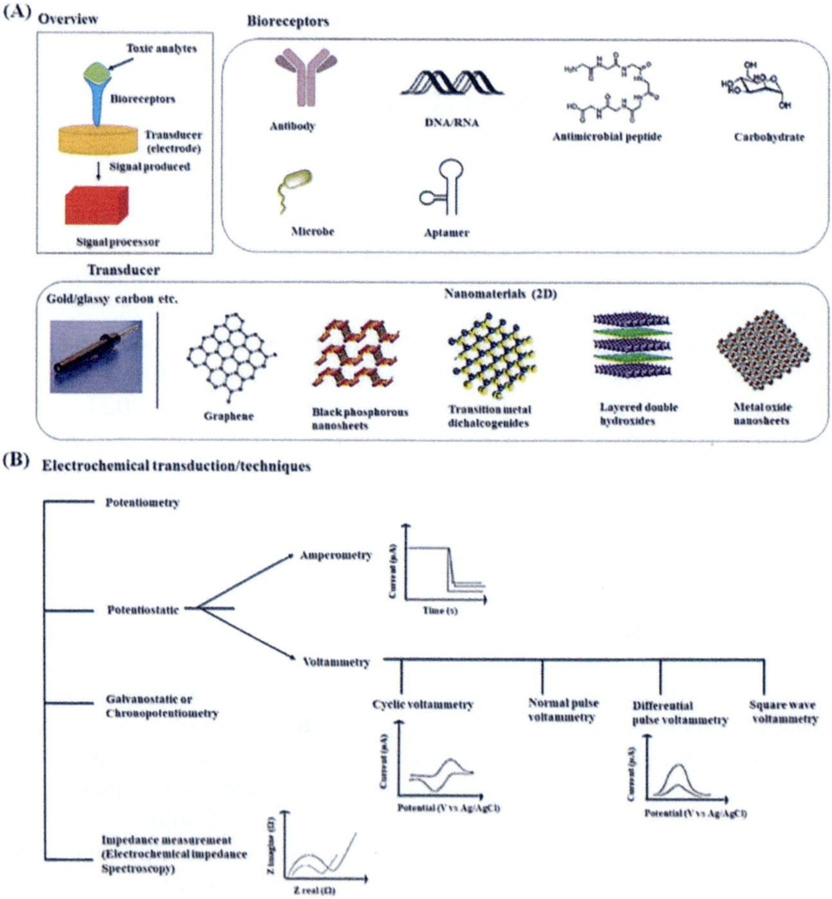

FIG. 6 (A) Schematic illustration of a typical electrochemical food sensor and its various components, including bioreceptors, transducer, and signal processor (Amiri et al., 2018). The structure of two-dimensional nanomaterials used to fabricate the electrode surfaces in biosensors has been shown (Hu et al., 2019). (B) Different types of electrochemical techniques have been presented (Raja et al., 2021; Vogiazi et al., 2019).

Nanostructured electrode surfaces made up of carbon nanomaterials and metal nanoparticles show promising results in the detection of food-borne allergens due to enhance electrochemical response, immobilization of reagents with extended lifetime, and strategizing no-wash electrochemical biosensors (Campuzano et al., 2019; Zhang and Chen, 2019). Immunosensors such as aptamers, peptides, nucleic acids, and oligonucleotide switching probes provide good alternatives for the detection of allergen (Li et al., 2018; Titoiu et al., 2019; Miranda-Castro et al., 2020). An aptamer-based biosensor employing graphene oxide/Au nanocomposite with hydrophobic channels on a filter paper was utilized for the detection of allergens arachin (Ara h 1) in peanuts, β-lactoglobulin (β LG) in milk, and tropomyosin (Pen a 1) for shrimp and other shellfish in a colorimetric manner. A nano-biosensor utilizing gold coating on screen-printed carbon electrode (SPCE) was utilized for the detection of Ara h 1 allergen in peanuts (for references Adriano Aquino and Conte-Junior, 2020). A nanoparticle quantum dot (QD) coupled with microfluidics and aptamer functionalized graphene oxide (GO) was utilized for the detection of Ara h 1 allergen (Weng and Neethirajan, 2016). Similarly, ELISA-based antibody-coated magnetic microparticles were utilized for the detection of Ara h ¾. A modified gold nanoparticle-polyethyleneimine-multiwalled carbon nanotube nanocomposite (AuNPs-PEI-MWCNTs) voltammetric immunobiosensor was fabricated for the label-free detection of kidney bean lectin (KBL) (Adriano Aquino and Conte-Junior, 2020). An electrochemical immunosensor fabricated on graphene-modified screen-printed-electrode (SPE) with immobilized β-lactoglobulin antibodies was used for the detection of β-lactoglobulin (Eissa and Zourob, 2017). LSPR based casein immunosensor with immobilized casein antibodies was used for the detection of casein in raw milk (Hiep et al., 2007).

Integrated exogenous antigen testing (iEAT) protocol was devised for the detection of the five most representative food antigens in peanuts, hazelnuts, wheat, milk, and eggs. The platform was tested for real-world samples showing antigen extraction and detection rapidly and cost-effectively (Lin et al., 2017).

Despite the significant development in the field of biosensing for food-borne allergens, much is required to be done due to the consistent emergence of novel allergens due to the extensive import of novel and nonnative food materials. This demands for the development of very specific sensors and bioreceptors. Biosensors need to be customized for the detection of different allergens in the same sample or same allergen in different samples simultaneously in a simple, reliable, and cost-effective manner. To enable robust food analysis, a biosensor fabrication strategy requires nanomaterials as electrode modifiers, nanocatalysts, nanoreporters, and nanocarriers. Additional features for improvisation may include minimal sample pre-treatment, reagent-less, calibration-free measurement by the biosensing platform (Idili et al., 2019; Li et al., 2018; Lin et al., 2017).

5.3.1 Microfluidics in food technology

Microfluidics integrates multiple technologies onto a single platform resulting in multiple miniaturized analytical processes representing a micro total analysis system (μTAS). Microfluidics offers the advantage of miniaturization, portability, and reduced sample and reagent volumes. However, the current challenges in microfluidics development are microfabrication techniques, electrokinetic and hydrodynamic flows, and electrochemical detection (Escarpa, 2014; López et al., 2016).

Microfluidics permits rapid, reliable detection of foodborne toxins, allergens, pathogens, toxic chemicals, heavy metals, and other contaminants. Microfluidics with its portability and minimized sample and reagent consumption makes it an ideal technology ensuring food safety. However, challenges remain in terms of complex food matrix preparation and complicated fabrication steps. Microfluidics-based sensing can be improvised by integrating biomolecules such as DNA and protein into the microfluidic system (Arduini et al., 2016; Weng and Neethirajan, 2017).

5.3.2 Microfluidics for the biomonitoring of food allergens

Microfluidics-based electrochemical sensors can be coupled with cell culture techniques to detect food-borne allergens by identifying the allergen-induced changes in cell morphology, and metabolism. A sensing platform was devised for the detection of food allergen, and dinitrophenylated bovine serum albumin (DNP-BSA). The sensing architecture consists of a microfluidic chip coupled with the gold electrode and the markers ANA-1 macrophages and RBL-2H3 mast cells, which changes in response to the food allergen DNP-BSA. The cellular allergic response was confirmed using enzyme-linked immunosorbent assay (ELISA) detecting inflammatory cytokines. This has validated the efficacy of the sensing platform in rapid, reliable, sensitive, selective manner facilitating real-time detection of cellular allergens (Jiang et al., 2016a, b).

5.4. Biomonitoring of food-borne pathogens

Food contamination can be direct when the pathogen is involved in the contamination process. It can be indirect when cross-contamination of food occurs due to different contaminated sources such as air, water, hair, dirt, animal feces, and infected wounds, etc. (Barbosa et al., 2019). Waterborne pathogens are the prime source of food contamination worldwide (Ali et al., 2020). Bacterial contamination of food constitutes 40% of the total infections. Pathogenic bacteria such as *Escherichia coli*, *Salmonella typhimurium*, *Staphylococcus aureus*, *Bacillus cereus* constitute 90% of food contamination worldwide (Mishra et al., 2018). *Streptococci* bacterial contamination leads to gut irritation, kidney affliction, cognitive incapacitation, joint inflammation, visual impairment, and death.

The toxins released by bacteria and fungi survive the process of food processing. The bacterial contamination of food requires monitoring at the food-chain level.

There are enumerable conventional assays and tests available for the detection of causative organisms such as viable cell counting, staining, carbohydrate fermentation assay, enzyme-linked immunosorbent assay, polymerase chain reaction, ultraviolet detection, and fluorescence techniques, etc. (for references Ali et al., 2020). However, these techniques suffer from the serious constraints of the requirement of trained and skilled manpower, sophisticated instrumental setup, time and labor-intensive assays, and affordability. Biosensors provide a promising alternative in being sensitive, selective, portable, and able to align with other bioanalytical devices for real-time detection of the pathogen in simple, reliable, and rapid manner. Bacterial antigens, toxins, microbially contaminated by-products, or spoilage precursors can be easily detected by biosensors (Ibrišimović et al., 2015).

Optical biosensors such as, Colorimetric, fluorescence, chemiluminescence, and surface plasmon resonance (SPR) identify bacteria in situ through refractive index measurement or through the thickness that develops as bacterial cells attach to the receptor on the transducer surface. A sandwich immunoassay can detect *Salmonella Typhimurium*, *Staphylococcus aureus*, *Salmonella enteritidis,* and *Campylobacter jejuni* in poultry. SPR biosensors have successfully detected several plants pathogens, viz. cucumber mosaic virus, *Pantoea stewartii*, plum pox virus, *Prunus necrotic ringspot virus*, citrus tristeza virus, and potato virus, cowpea mosaic virus, tobacco mosaic virus, lettuce mosaic virus, *Fusarium culmorum*, *Phytophthora infestans*, and *Puccinia striiformis* (for references Ali et al., 2020).

Electrochemical biosensors based on impedimetric, potentiometric, amperometric, electrochemiluminescent, voltammetric, and conductometric methods can detect genetically modified organisms (GMOs) such as *Yersinia enterocolitica, Salmonella Typhimurium* (Silva et al., 2019) (Fig. 7).

The **mechanical biosensors** can be of different types based on sensor-analyte interaction, viz. affinity-based assays, fingerprint assays, separation-based assays, and spectrometric assays (Arlett et al., 2011). A Quartz crystal microbalance (QCM)-aptasensor can rapidly detect *Brucella melitensis* in milk and milk products; *Campylobacter jejuni*, and red wine mycotoxin (ochratoxin), etc.

An EIS-based biosensor employing AuNPs and GO were utilized for the detection of the most common food pathogen, *Salmonella typhimurium* from pork or spiked food samples using poly [pyrrole-*co*-3-carboxyl-pyrrole] copolymer-supported aptamer-based impedimetric biosensor. *Staphylococcus aureus* from pigskin was detected using single-walled carbon nanotubes (SWCNT) based potentiometric sensor (Figs. 8 and 9). An electrochemical genosensor utilizing a modified pencil graphite electrode was used to determine *Bacillus cereus* from milk and infant formula and *Aeromonas hydrophylla* from fish and vegetables (for references Mishra et al., 2018). A colorimetric biosensor coupled

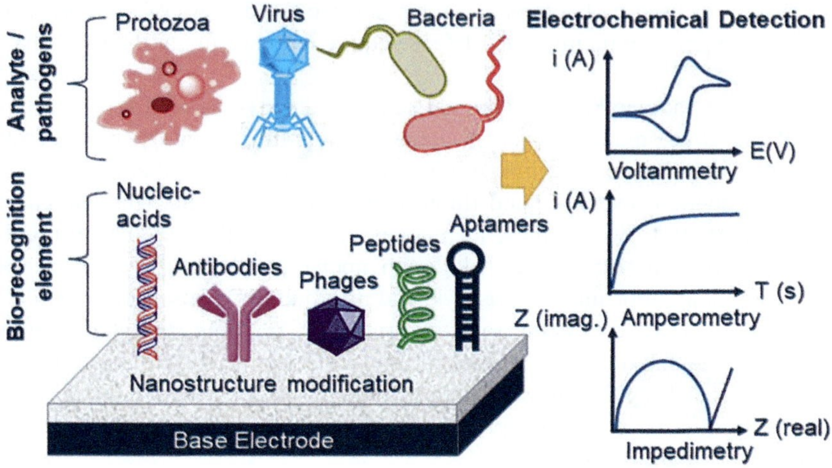

FIG. 7 Schematic representation of the key components of a biosensor platform for the detection of foodborne pathogens (Bwambok et al., 2020). *Copyright © 2020 by the authors. Licensee MDPI, Basel, Switzerland. This article is an open-access article distributed under the terms and conditions of the Creative Commons Attribution (CC BY) license (http://creativecommons.org/licenses/by/4.0/).*

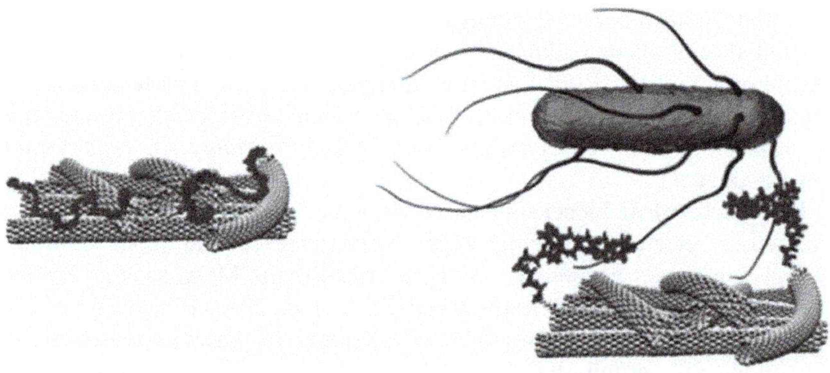

(a) (b)

FIG. 8 (A) Possible conformations of the aptamers that are self-assembled on carbon nanotubes; (B) Schematic representation of the interaction between the target bacteria and the hybrid aptamer-SWCNT system (Zelada-Guillén et al., 2009; Mustafa et al., 2017). *Copyright © 2017 by the authors. Licensee MDPI, Basel, Switzerland. This article is an open-access article distributed under the terms and conditions of the Creative Commons Attribution (CC BY) license (http://creativecommons.org/licenses/by/4.0/).*

with peptides immobilized on a gold sensor chip was utilized for the detection of *Listeria monocytogenes* from milk and meat (Alhogail et al., 2016). A 3D-nanorod based sensor and others based on surface-enhanced Raman scattering (SERS) was used to kill or deactivate bacteria (Wang et al., 2015; Yang et al., 2017).

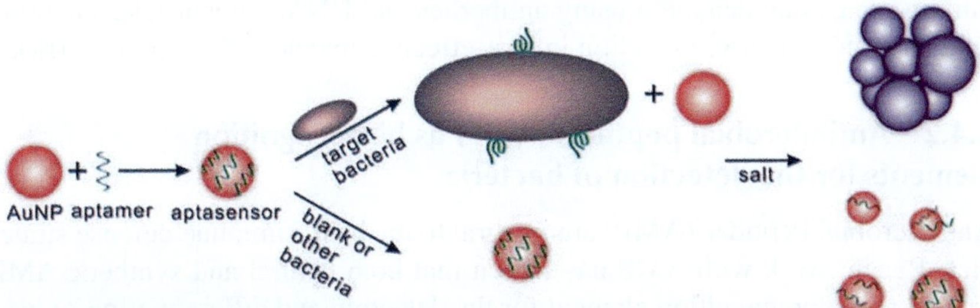

FIG. 9 Colorimetric detection of bacteria using aptamers and AuNPs (Mustafa et al., 2017). *Copyright © 2017 by the authors. Licensee MDPI, Basel, Switzerland. This article is an open-access article distributed under the terms and conditions of the Creative Commons Attribution (CC BY) license (http://creativecommons.org/licenses/by/4.0/).*

Adenosine triphosphate (ATP) based bioluminescence biosensors are reliably used for the hazard analysis and critical control point (HACCP) program for in situ hygiene monitoring by the identification of various food-borne microbes. Bioluminescence-based identification of bacterial ATP is a powerful indicator of bacterial contamination in food, poultry, dairy products, and cross-contamination of surfaces with minimum sample consumption as compared to the conventional assays. It provides a real-time picture of the contamination through physiologically relevant data. It is a simple, reliable, rapid, economical, less labor-intensive, and instrument-specific technique for microbial monitoring (Griffiths, 1996; Mandal and Biswas, 2020).

A bioluminescence-based ATP assay utilizing anti-bacterial peptide-coated magnetic spheres has successfully isolated gram-positive bacteria from the gram-negative bacteria (Fan et al., 2020). Approximately, 1 picogram of ATP is equivalent to 1000 bacterial cells in the sample. A novel adenylate (A3) based assay was developed to monitor hygiene for the cases where ATP-based assays are not effective (Bakke and Suzuki, 2018).

Radiofrequency identification (RFID) based sensor can reliably determine the presence of *Escherichia coli* (*E. coli*) and *Salmonella* in packaged food materials. This system utilizes immobilized toxins on RFID tags incorporated within the package. This automated sensing platform is connected with a wireless system to release output used for the real-time monitoring of foodborne pathogens in the packaged food materials (Mustafa and Andreescu, 2018). A LIG biosensor having a laser-guided graphene electrode functionalized with antibody was used for the electrochemical detection of *Salmonella enterica*, a causative organism of major foodborne illness worldwide (Soares et al., 2020).

5.4.1 *Salmonella* spp.

Salmonella spp. is one of the most common pathogens responsible for the contamination of food and outbreak of food-related diseases worldwide. Biosensors for *Salmonella* spp. detection employs phase-based separation and immuno-magnetic detection using antibodies and DNA nanoparticles enabling simple, rapid, sensitive detection in cost-effective manner (Silva et al., 2018).

5.4.2 Antimicrobial peptides (AMP) as biorecognition elements for the detection of bacteria

Antimicrobial Peptides (AMP) are integral to the host's immune defense strategies. Recent work with AMP has shown that both natural and synthetic AMP can act as a biorecognition element for the detection and differentiation of bacteria with high stability and activity. A supramolecular AMP synthesized artificially was used to fabricate a sensor using AuNPs-functionalized electrode modified with AMP. The platform showed antimicrobial activity against *E. coli*, *Pseudomonas aeruginosa* (*P. aeruginosa*), *Staphylococcus aureus* (*S. aureus*), and *Staphylococcus epidermidis* (*S. epidermidis*). Electrochemical impedance spectroscopy was used as a transducer to quantify the binding of bacteria enabling rapid detection in the label-free manner (Liu et al., 2016a, b). A colorimetric assay was developed using a magnetic nanoparticle (NPs)-protease-gold sensing probe for the detection of *Listeria monocytogens.* The assay was successfully utilized with high specificity against important food-borne bacteria such as *E. coli*, *Salmonella*, *Shigella flexnerii*, and *Staphylococcus aureus* (Alhogail et al., 2016).

A SERS-based multifunctional chip made of a silicon wafer containing AgNPs modified with 4-mercaptophenylboronic acid (4-MPBA) was used for the simultaneous detection of *E. coli* and *S. aureus.* It has inactivated the pathogen in blood by the action of Ag^+ released from AgNPs (Wang et al., 2015). Aptamer functionalized Fe_3O_4 magnetic NPs were used to separate and isolate *Staphylococcus aureus*, *Vibrio parahemolyticus*, and *Salmonella typhimurium* from solution (Wu et al., 2014). Similarly, FeCoNPs were used to separate phage-bound bacteria from the unbound bacteria (Chen, 2015). In another assay, *Bacillus cereus* and *E. coli* were detected and separated by the screen-printed-carbon electrode (SPCE) utilizing electrochemical detection and immunomagnetic separation (Setterington and Alocilja, 2012). Vancomycin-functionalized AgNPs/3D-ZnO nanorod arrays were utilized for the simultaneous detection, inactivation, and elimination of pathogens (Yang et al., 2017).

5.4.3 Molecularly imprinted polymers (MIP) for biomonitoring of food-borne bacteria

Molecularly Imprinted Polymers (MIP) are artificial recognition elements with high chemical and mechanical stability, low cost, and easy synthesis. This makes them advantageous over natural recognition elements. MIP leads to the creation of specific recognition sites by the polymerization of monomers in the presence of template molecules. Removal of template creates a "shape memory cavity" with binding ability that will serve as a recognition site for the molecule with identical geometry monitored by SPR and QCM based sensors (Mustafa et al., 2017). A QCM-based MIP bacterial platform was fabricated employing electrochemically polymerized polypyrrole (PPy) deposited as a thin film on a gold-evaporated quartz crystal. This method was applied successfully with *Pseudomonas aeruginosa, Acinetobacter calcoaceticus, E. coli*, and *Serratia marcescens* in apple juice samples (Tokonami et al., 2013). In another study, a MIP-based QCM and SPR assisted detection of *E. coli* in an apple juice sample was achieved. The sensing platform involved modification of QCM and SPR gold surfaces with allyl mercaptan and *N*-methacryloyl-L-histidine methylester monomers having similarities with natural antibodies. UV-photo polymerization facilitated micro-contact imprinting of *E. coli* (Yilmaz et al., 2015). A MIP sensor fabricated with 2,5-dimethyl-4-hydroxy-3(2*H*)-furanone (DMHF) was used to detect gram-negative bacteria such as *Aeromonas hydrophila* and *Pseudomonas aeruginosa* using their quorum signaling molecules, such as *N*-acyl-homoserine-lactones (AHLs) (Jiang et al., 2016a, b).

5.4.4 Bacteriophage

Bacteriophages are known to parasitize and lyse bacterial cells. Phages can be reliably used for the identification of bacteria and differentiation of bacterial strains. Phages show strong specificity for their host, hence can be used for differentiating live cells from the dead. These properties in conjunction with low cost make phages a good choice for the molecular recognition events through biosensors for the detection of bacteria (Mustafa et al., 2017). A multifunctional T7 bacteriophage was designed for the detection, isolation, and concentration of *Escherichia coli (E. coli)* from drinking water using colorimetric signal analyzed by UV spectrometry (Chen et al., 2015). The phage-based magnetic probe was successfully used against *Salmonella enterica (S. enterica), Staphylococcus aureus (S. aureus)*, and *Pseudomonas aeruginosa (P. aeruginosa)* (Fig. 10) (Derda et al., 2013). A biosensor for the detection of *Salmonella typhimurium* was fabricated. It involved adsorption of bacteriophage on a piezoelectric transducer leading to decrease in resonance frequency allowing its quantification (Olsen et al., 2006). A QCM with dissipation (QCM-D) based detection and differentiation platform was designed for the detection and differentiation of methicillin-resistant (MRSA) and sensitive (MSSA) *Staphylococcal* species by immobilizing lytic phages on the QCM sensor. As discussed earlier, the binding of bacteria with phage resulted in the reduced resonance frequency and increased dissipation energy. MRSA and MSSA were differentiated owing to their ability to bind with penicillin-binding protein antibodies. MRSA binds to the antibody due to its high specificity while MSSA did not (Tawil et al., 2012). Other bacteriophage-based sensing strategies involve quantification of lysis products by amperometry or impedimetric sensors. For example, the presence of β-D-Galactosidase led to the quantification of *E. coli*, where p-aminophenyl-β-D-galactopyranoside (β-PAPG) was used as a substrate to detect p-aminophenol

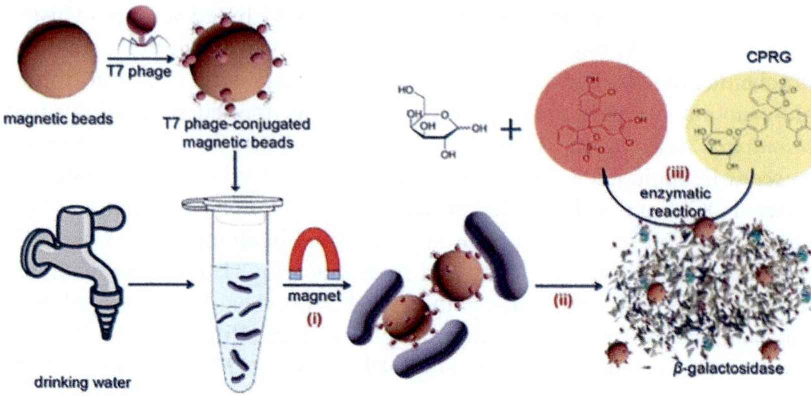

FIG. 10 Schematic representation of the use of T7 bacteriophage-conjugated magnetic probe to detect *Escherichia coli* in drinking water (i) Introduction of T7 bacteriophage-conjugated magnetic probe to attack *E. coli* and separate it by the influence of magnet (ii) The explosion or lysis of *E. coli* and the release of T7 phages and β-gal; (iii) β-gal catalyzed CPRG hydrolysis to produce a colorimetric signal (Mustafa et al., 2017). *Copyright © 2017 by the authors. Licensee MDPI, Basel, Switzerland. This article is an open-access article distributed under the terms and conditions of the Creative Commons Attribution (CC BY) license (http://creativecommons.org/licenses/by/4.0/).*

through oxidation at carbon electrode where impedance measurement of bacteria was done without quantifying the lysis product (Neufeld et al., 2003). T4 bacteriophage was directly immobilized on screen-printed-carbon-electrode (SPCE) microarrays using magnetic beads to act as a specific probe. Impedance spectroscopy was used to quantify the bacteria captured by the bacteriophage (Shabani et al., 2013) (Fig. 11).

5.5. Biomonitoring of food-toxins

Food toxins are the metabolic by-products generated by fungi, algae, bacteria, and plants. Mycotoxins or the fungal toxins are secondary metabolites of fungus of which more than 400 types are already reported. Food infested with fungi shows the presence of mycotoxins and other toxic metabolites leading to serious health complications culminating in cancer. These toxins show a great diversity in their chemical structure, physicochemical properties, and toxicological behavior (Marin et al., 2013). Toxins can be carcinogenic, teratogenic, or allergenic. Detection of these toxins is tricky as these are present in trace amount and occur in various combinations either produced by single fungal species or multiple fungal species. A folding mechanism-based electrochemical biosensor utilizing MB-tagged anti-OTA aptamers was used for the determination of ochratoxin A (OTA) (Mishra et al., 2015). In another work, aflatoxin M1 (AFM1) was identified in raw milk by impedimetric biosensor using hexaethylene glycol-modified aptamer on screen-printed-electrode (SPE) (Istamboulié et al., 2016) (Fig. 12).

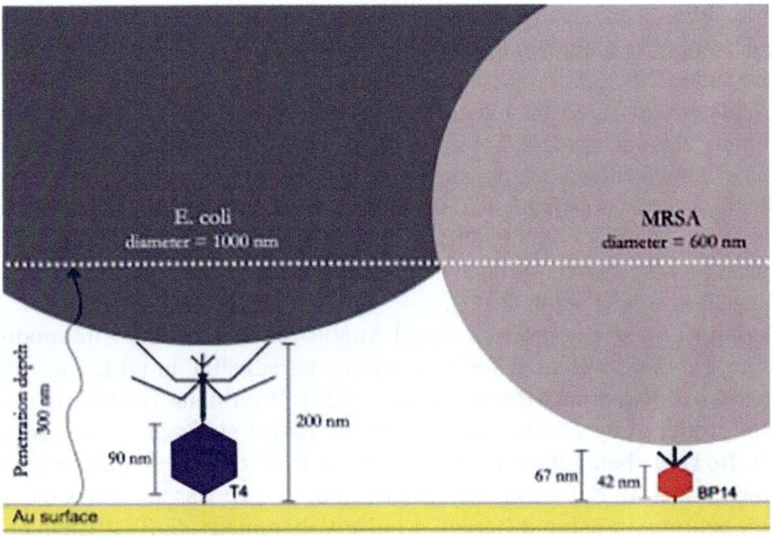

FIG. 11 Concept of a bacteriophage-based sensor for *E. coli* and MRSA using covalently attached T4 and BP14 bacteriophages (Mustafa et al., 2017). *Copyright © 2017 by the authors. Licensee MDPI, Basel, Switzerland. This article is an open-access article distributed under the terms and conditions of the Creative Commons Attribution (CC BY) license (http://creativecommons.org/licenses/by/4.0/).*

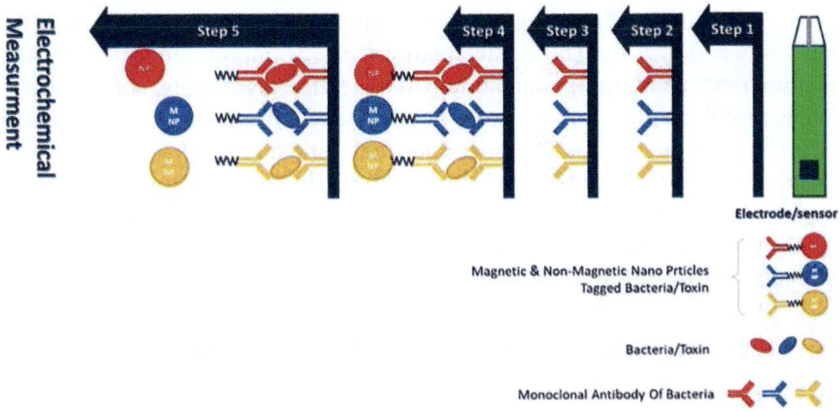

FIG. 12 Schematics for bacteria and toxin detection using antibody-based electrochemical biosensor (Mishra et al., 2018). *Copyright © 2018 by the authors. Licensee MDPI, Basel, Switzerland. This article is an open-access article distributed under the terms and conditions of the Creative Commons Attribution (CC BY) license (http://creativecommons.org/licenses/by/4.0/).*

Bioreceptor arrays contain electrodes functionalized with different bioreceptors with specific binding targets. Optical and Piezoelectric sensors are employed for the detection of toxins and chemicals. Fluorescent nanoparticles are developed to detect surface-, inter- and intracellular toxins (Saucedo and Mulchandani, 2016; Patra et al., 2017; Bahadır and Sezgintürk, 2017).

Toxin separation and quantification from food can be achieved by various sensing platforms in conjunction with (SERS) (Moran et al., 2016; Liao and Lu, 2016; Stadler, 2016).

Okadaic acid (OA) is a major marine lipophilic biotoxin produced by Dinoflagellates such as *Dinophysis* and *Prorocentrum*. Okadaic acid (OA) contaminated shellfish lead to a physiological and pathological condition known as diarrhetic shellfish poisoning that includes gastronomical disturbances (Callejas et al., 2015; Eangoor et al., 2017). A voltammeter immunosensor having graphene-modified screen-printed electrode (GSPE) detected OA using enzyme labeling (Eissa and Zourob, 2012).

An impedimetric aptasensor based AuNPs/rGO nanocomposite modified graphite electrode was used for determining tetracycline in milk, meat, fish, and shrimp (Mohammad-Razdari et al., 2020). Botulinum (BoNT) is a neurotoxin produced by *Clostridium botulinum*. BoTN exists in seven serotypes (A-G), BoTN/A being the most severe. It can lead to severe paralysis even in the lower dose. An rGO/Au electrode-based biosensor was used to detect botulinum neurotoxin using differential pulse voltammetry (DPV) (Chan et al., 2015).

5.5.1 Microbial fuel cells for food safety and security

The microbial fuel cells can be used for monitoring toxicity and biochemical oxygen demand (BOD) in water, thus serving as an early warning system for food safety and security. The sensing platform can be improvised for sensitivity, detection range, and limit, integrating with detection algorithms, and empirical models to differentiate toxicity from BOD on account of simultaneous occurrence (Jiang et al., 2018). The microbial fuel generation cells relying on microbial interaction can generate electricity through the oxidation of organic or inorganic wastes. Biofuel production from the combination of plants, microbes, and algae can provide a green and environment-friendly alternative to power and energy in the future (Pinto and Ferranti, 2016).

5.6. Quality control in the food and beverage industry

The food and beverage industry relies heavily on the quality of its products as per the standard norms. Nanotechnology-based monitoring ensures structure, composition, nutrient, and microbial composition of the products with high sensitivity, specificity, and with a lower limit of detection.

Chitosan-nanoparticles are nontoxic, biodegradable, biocompatible with good adsorption ability. It encapsulates vital agrochemicals and genetic materials for their slow and regulated release to the plants. However, challenges remain in terms of imparting knowledge of encapsulation, scaling up the production, lowering the cost of application, and toxicology for realizing its potential (Kashyap et al., 2015).

5.6.1 Food-fingerprinting (foodomics)

Food fingerprinting or foodomics is a fast-emerging field in food technology dealing with the nutritional value, quality, authenticity, safety, and security of food. Food fingerprints in different food can be determined using analytical techniques like phytochemistry, food chemistry, bioinformatics, and biosensing technologies. These techniques are expected to identify food-related novel biomarkers to ensure food quality and authenticity (Gallo and Ferranti, 2016).

5.6.2 Dairy products

NIR spectroscopy-based scanning monochromator (SMC) was utilized for the classification of cheese from the milk mixture of cow, ewe, and goat, and SMC spectrometer to quantify the water added in the milk (Kamboj et al., 2020; Soto-Barajas et al., 2013). FT-IR based on the NIR module was used for the determination of adulterants such as micro-cellulose, silicon dioxide, wheat flour, wheat semolina, and sawdust in grated cheese (Visconti et al., 2020).

5.6.3 Sweeteners

NIR coupled with Raman spectroscopy was used for the determination of the p-hydroxybenzoic acid (PHBA) content of honey (Tahir et al., 2019). An FT-NIR spectrophotometer was used for the analysis of adulteration in honey using glucose/fructose solutions (Zhu et al., 2010). Benchtop spectrometer with aquaphotomics detected the adulteration of honey with HFCS or rice, beet, corn, maltose, high-fructose corn syrup, and sucrose syrup using TR-FTIR spectrometer (Bázár et al., 2016; Huang et al., 2020).

5.6.4 Beverages: Coffee and tea

A benchtop NIR spectrometer was used to assess the genotype and geographical origin of coffee based on spectral patterns (Marquetti et al., 2016). FT-NIR was used for characterizing the coffee in terms of color and arabica content or classifying coffee into *robusta* and *arabica* using ATR-FTIR (Bertone et al., 2016; Craig et al., 2018). NIR portable devices were used for determining the catechin, polyphenol, and amino acid content of instant green tea (Sun et al., 2020; Wang et al., 2019).

5.6.5 Fruit juices

Citric and tartaric acids were quantified using a reflectance mode SMC spectrometer (De Carvalho et al., 2015). SMC NIR in reflectance model was used to classify and discriminate green and orange melon varieties (Németh et al., 2019). Transmittance mode benchtop FT-NIR was used to measure total yeast assimilable nitrogen, free amino nitrogen, and ammonia in grape juice samples over different seasons (Petrovic et al., 2020).

5.6.6 Soft drinks

Benchtop SMC spectrometer was used to quantify glucose, fructose, and sucrose content of commercial soft drinks or the caffeine level in the soft drinks using transmission mode FT-NIR spectrometer (Ilaslan et al., 2015; Li et al., 2007). Benchtop FT-NIR can also be used for measuring the ascorbic acid content of the powdered soft drinks. This is an advantage over titration and chromatographic methods, which are expensive and time-consuming (De Santana et al., 2020). An amperometric glucose biosensor based on GOx immobilization with polymeric hydrogel (chitosan) on titanium dioxide nanotube arrays (TiO_2NTAs) was fabricated for the detection and quantification of glucose in different food products such as soft drinks, soy sauces, dairy products, and tomato sauces. The Chitosan-GOx/TiO2NTAs/Ti biosensor quantified the accurate glucose concentration in a rapid, selective, and economical manner. The results obtained were in agreement with other analytical results (Artigues et al., 2017).

5.6.7 Meat

The noninvasiveness of NIR spectroscopy has multiple advantages for the assessment of meat quality. Benchtop SMC spectrometer can determine protein and intramolecular fat content of rabbit hind meat or fatty acid composition from steers fed with sunflower and flax seeds, or total lipid content of beef (Bazar et al., 2007; Mourot et al., 2015; Prieto et al., 2014). NIR spectroscopy can be used to determine sodium content in commercially processed meat, vacuum dried ham slices, and ash and dry matter in freeze-dried ostrich meat (Campos et al., 2017; De Marchi et al., 2017; Viljoen et al., 2005). NIR spectroscopy with bifurcated optical cable can accurately determine tenderness and juiciness in pork (Zamora-Rojas et al., 2013). Benchtop SMC NIR spectrometer can detect the adulteration of minced beef with turkey meat, minced lamb meat with pork, minced llama meat with horse meat, and vegetable protein, and soy flour in minced pork and beef (Alamprese et al., 2013; Kamruzzaman et al., 2013; Mamani-linares et al., 2012; Rady and Adedeji, 2018).

5.7. Biomonitoring of antibiotics for food safety

Human consumption of food products such as milk, egg, meat, and chicken contaminated with antibiotics used as antibacterial and growth enhancers will lead to drug resistance in the human body. The notable antibiotics present in the food and farm products are chloramphenicol, tetracycline, erythromycin, tetracycline, penicillin, ampicillin, quinolone, and cephalosporins. These antibiotics are used injudiciously to enhance growth and muscle mass. The result is health problems to humans when these antibiotics reach to humans through the food-chain via the process of biomagnification. It also leads to the development of antibiotic-resistant food-borne pathogens (Gaudin, 2017). To circumvent the issues of drug resistance of animal origin and toxicity of pathogens, a "Precautionary principle" model is enforced with a "maximum residue limit" by regulatory authorities. It was defined as "the maximum legally acceptable quantity of pharmacologically active substances (whether active principles, excipients or metabolites) and their degradation products in food products derived from animals". It depends on the "acceptable daily intake" (ADI), which takes into account the maximum safety margin. In the backdrop of this, it is imperative to have a simple, reliable, sensitive, and selective screening assay/protocol which can monitor the veterinary drug residues in animal-derived food (Majdinasab et al., 2017).

To monitor the presence of veterinary drug residues, immunological sensors based on antigen-antibody affinity pairs are the most appropriate. These affinity biosensors use optical transducers while some studies have utilized electrochemical transducers as well (Majdinasab et al., 2017). Electrochemical biosensors have widespread use in the field of disease diagnosis, environmental monitoring, food safety, etc. The incorporation of nanomaterials such as metal nanomaterials, silica nanoparticles, carbon nanomaterials, and other functionalized nanoparticles enhance the selectivity, sensitivity, and speed for sensing antimicrobial drug residues in animal-derived food (Majdinasab et al., 2017).

5.7.1 Kanamycin (KAN)

Kanamycin (KAN) is an important subclass of aminoglycoside antibiotics produced by the fermentation of *Streptomyces kanamyceticus*. An amperometric immunosensor based on graphene sheet-Nafion-thionine-platinum nanoparticles (GS/Nf/TH/Pt)-modified electrode was used for the ultrasensitive detection of kanamycin in the label-free manner (Wei et al., 2012). An immunosensor was fabricated with silver hybridized mesoporous ferroferric oxide nanoparticles (Ag@Fe_3O_4 NPs) and thionine-mixed graphene sheet (TH-GS) for the detection of kanamycin in pork meat samples in a label-free manner (Yu et al., 2013). An aptasensor designed by the self-assembly of gold (AuNPs)/conducting polymer (2,5-di-(2-thienyl)-1*H*-pyrrol-1-(*p*-benzoic acid)) nanocomposite on screen-printed-electrode (SPE) was utilized for the detection of kanamycin in food samples (Zhu et al., 2012). Electrochemical impedance spectroscopy (EIS) based aptasensor on the assembly of single-strand DNA (ssDNA) anti-KANA-aptamer-functionalized screen-printed-carbon electrodes (SPCE) was utilized for the quantitative detection of kanamycin in spiked milk samples (Sharma et al., 2017). Other types of biosensors utilized for the sensing of kanamycin are square wave voltammetry (SWV) based immunosensor and aptasensor, photoelectrochemical aptasensor, and differential pulse voltammetry (DPV) based aptasensor, etc. (Majdinasab et al., 2017).

5.7.2 Chloramphenicol (CAP)

Chloramphenicol (CAP) is a broad-spectrum antibiotic used for the treatment of infections in humans and animals. A novel electrochemical biosensor based on multiwalled carbon nanotubes@molecularly imprinted polymer (MWCNTs@MIP) was used for sensing chloramphenicol in spiked milk and honey samples. MWCNTs@MIP was used to coat glass carbon electrodes (GCE) modified with mesoporous carbon and 3-dimensional porous graphene to enhance the sensitivity of the biosensor (Yang and Zhao, 2015). An electrochemical biosensor was fabricated based on monoclonal antibody immobilized onto hollow gold nanospheres/chitosan composite coating on the surface of a glassy carbon electrode (GCE). Differential pulse voltammetry was used for the quantification of meat, fish, and pork samples spiked with three different concentrations of chloramphenicol (Zhang et al., 2011). An electrochemical aptasensor based on CAP-aptamer coated onto the surface of gold electrode was utilized for the detection of chloramphenicol via differential pulse voltammetry from the spiked honey samples (Yan et al., 2012). An electrochemical sensor based on 3D-reduced graphene oxide architecture (3DRGOA) coated onto the surface of glassy carbon electrode (GCE) was utilized for the detection of chloramphenicol through differential pulse voltammetry (DPV) (Zhang et al., 2016). An electrochemical sensor based on Molybdenum disulfide (MoS_2)/self-doped-polyaniline (SPAN) nanocomposite was fabricated. The resulting nanocomposite (MoS_2/SPAN) was used to coat carbon-paste-electrode (CPE). This platform was used for the detection of conjugate structured CAP (Yang and Zhao, 2015).

An electrochemical sensor was fabricated for the simultaneous detection of chloramphenicol and oxytetracycline (OTC). The sensing platform was based on high-capacity magnetic hollow porous (MHP) nano-tracer coupling exonuclease-assisted target recycling. The MHP was utilized for the immobilization of Cd^{2+} and Pb^{2+} to enhance the signal and simplify the separation and detection process. Metal-ion immobilized MHP were coupled with DNA strands. This was followed by the immobilization of CAP and OTC on complementary strands and subsequent immobilization on glass-carbon-electrode (GCE). The CAP and OTC were released from the DNA strands after incubation for another round of recycling leading to signal enhancement (Yan et al., 2016). Another multiplex detection assay was designed based on nanoscale metal-organic frameworks (NMOF) as a substrate, and circular strand-replacement DNA polymerization (CSRP) as a target for the simultaneous detection of CAP and OTC (Chen et al., 2017a, b).

Other sensors employed for the detection of chloramphenicol are differential pulse voltammetry based molecularly imprinted sensor, differential pulse voltammetry (DPV) based aptasensor and immunosensor, cyclic voltammetry (CV) and linear sweep voltammetry (LSV) based aptasensor, square wave voltammetry (SWV) based aptasensor, amperometric sensor, and potentiometric based molecularly imprinted sensor, etc. (Majdinasab et al., 2017).

5.7.3 Tetracycline

An electrochemical sensor based on platinum-catalyzed hydrogen evolution reaction (HER) was developed for the detection of tetracycline in spiked milk, honey, and peanut samples. The sensing output was based on the competitive interaction between the tetracycline as a target and TC-BSA labeled with GN-PtNPs to the anti-TC antibody immobilized onto the surface of a gold electrode (Que et al., 2013a, b). An M-shaped electrochemical aptasensor based on potassium hexacyanoferrate (III) (K3[Fe(CN)6])/potassium hexacyanoferrate (II) trihydrate (K4[Fe(CN)6]·3H2O) was used as a redox probe for the measurement of differential pulse voltammetry (DPV) and cyclic voltammetry (CV) output signal for the ultrasensitive detection of tetracycline (Taghdisi et al., 2016). Another electrochemical aptasensor having a glassy carbon electrode (GCE) modified with multi-walled-carbon-nanotube (MWCNTs) was fabricated for the highly sensitive detection of tetracycline in milk samples (Zhou et al., 2012). Molecularly imprinted polymers (MIP) are crucial for the development of artificial recognition elements. A gold nanoparticle-based MIP electrochemical sensor was utilized for the detection of tetracycline in spiked honey samples. A nanocomposite with reduced graphene oxide (rGO)-magnetic nanoparticles (Fe_3O_4)/sodium alginate (SA) was used to coat screen-printed-carbon-electrode (SPCE). The rGO-Fe_3O_4/SA nanocomposite and tetracycline aptamer were immobilized on the modified surface for the rapid, sensitive, and selective determination of tetracycline (Zhan et al., 2016).

Other sensors employed for the detection of tetracycline are cyclic voltammetry (CV) based immunosensor, differential pulse voltammetry (DPV) based immunosensor, differential pulse voltammetry (DPV) based aptasensor, linear sweep voltammetry (LSV) based molecularly imprinted (MIP) sensor, electrochemical impedance spectroscopy (EIS) based aptasensor, cyclic voltammetry (CV), and square wave voltammetry (SWV) based aptasensor, amperometric immunosensor, etc. (Majdinasab et al., 2017).

5.7.4 Streptomycin (STR)

An electrochemical biosensor based on molecularly imprinted polymer (MIP) synthesized through the copolymerization of aniline and *o*-phenylenediamine on the surface of gold electrode was fabricated for the detection of tetracycline in honey and spiked milk samples (Que et al., 2013a, b). An electrochemical aptasensor based on gold nanoparticle-functionalized magnetic multi-walled carbon nanotubes (GNP-MWCNTs-Fe_3O_4) and nanoporous PtTi (NP-PtTi) alloy was utilized for the detection of streptomycin. The GNP-MWCNTs-Fe_3O_4 nanocomposite was used to coat glassy-carbon-electrode (GCE) followed by the immobilization of STR-aptamer on NP-PtTi/GNP-MWCNTs-Fe_3O_4/GCE. This sensing platform showed improved sensitivity under optimized experimental conditions (Yin et al., 2016a, b). Another electrochemical sensor-based "signal attenuation" was fabricated for the sensitive detection of streptomycin. The sensing platform was composed of MWCNTs-CuO-AuNPs nanocomposite used to coat glassy-carbon-electrode (GCE) followed by casting the resulting suspension on porous carbon nanorods (PCNRs) modified electrode. The resulting electrode was functionalized with streptomycin aptamer leading to a complex formation between streptomycin and immobilized aptamer and generation of the output signal (Yin et al., 2016a, b). A biosensor based on a gold electrode, streptomycin aptamer, complementary strand (CS) of aptamer, exonuclease I (Exo I), and ferricyanide as redox probe was used to quantify streptomycin in milk samples utilizing differential pulse voltammetry (DPV) (Danesh et al., 2016). Another study has developed a simple immunoassay based on the competition between STR analyte and STR-BSA conjugates for the sensitive determination of streptomycin (Liu et al., 2011).

5.8. Biomonitoring of spurious chemicals during food processing

Food processing is not safe entirely. Some dangerous contaminants such as acrylamide, acrolein, advanced glycation end products, chloropropanols, glycidols, furan, heterocyclic aromatic amines, hydroxyl methyl furfuryl, and polycyclic aromatic amines arise during various steps of food processing. The fermentation process yields acetaldehyde, biogenic amines, and ethyl carbamate. The preservatives and additives added to increase the shelf-life and quality of food products sometimes get transform into undesirable benzene, methylimidazoles, nitrosamines, and semicarbazide, which are unfit for human consumption. The alternative techniques developed to ward off the disadvantages of thermal processing produces alkylcyclobitanones, free radicals, and furan-like highly toxic compounds (Stadler et al., 2002; Batra et al., 2013).

5.8.1 Biogenic amines monitoring as an indicator for food quality and safety

Biogenic amines, such as histamine, tyramine, cadaverine show a direct link with food quality, safety, and wastage. High levels of biogenic amines lead to food poisoning whereas low level leads to food intolerance. Fermented food items like cheese, wine, sausage, and pickled vegetables contain histamine, cadaverine, tyramine, and/or putrescine while fermented fishes have high levels of biogenic amines. Both cases show a high level of microbial load and microbial contamination (Spano et al., 2010; Doeun et al., 2017). Therefore, the total concentration of biogenic amines acts as an indicator of microbial contamination, and hence the indicator of quality and safety indices. The European Union has fixed the histamine threshold at 100–200 mg/kg while FDA, USA has fixed it at 500 mg/kg beyond, which food is not considered safe for consumption (U.S. Food and Drug Administration (USFDA), 2011; Visciano et al., 2014). A large chunk of fermented food and fish markets are located around the sea-coast in the developing or under-developed countries, where access to laboratory-based infrastructural setup and the analytical cost is expensive and unaffordable leading to the avoidance of the BA detection process. Mostly, the electrochemical biosensor for the biogenic amines utilizes diamine oxidase (DAO) as a marker as it selectively oxidizes biogenic amines (Vanegas et al., 2017). Therefore, sensors based on DAO are available. However, nanocarbon-based biosensors (graphene, carbon nanotube) offer great promise. An amperometric biosensing platform was designed using graphene functionalized with DAO and copper microparticles for the detection of biogenic amines from fish paste samples fermented with lactic acid bacteria. Results show that the sensor performs selectively in a complex food matrix in a low cost and rapid manner (Vanegas et al., 2018).

5.9. Nanomaterial toxicity: Biosensing in food quality assurance

The most pertinent issue in the use of biosensing for food quality assurance is the potential toxicity of the nanomaterials. A nontoxic, organic biosensing material-based bio-organic sensor can provide smart sensing solutions. Graphene-based nanomaterials owing to their superior optical, electrical, thermal, mechanical, and chemical absorption, anti-bacterial, anti-functionalization properties make them a favorite candidate for food industry application (Sundramoorthy and Gunasekaran, 2014). Electro-chemiluminescence biosensors coupled with nanomaterials, aptamers, immunoassays, and microfluidics were applied for the food quality measures in pesticides, toxins, residual drugs, heavy metals, bacteria, and viruses (Neethirajan et al., 2018a, b). Microfluidics integrated with aptamers, and SERS can detect food-borne allergens, pathogens, residual pesticides, heavy metals, toxins, additives, and other contaminants (Weng and Neethirajan, 2017). Nanoproteomics combined with SERS can provide protein information in a label-free fashion with high sensitivity, and portability. A laser ranging and laser scanning sensor with a remote sensing facility was fabricated for the measurement of crop vitality. A soil biosensor with the nano-lipid platform was designed for the detection of the nitrite content of the soil. Biosensors can also provide real-time information on the live food material such as fish by quantifying stress biomarkers such as glucose. Robotics, combined with biosensors and drones can provide real-time information on soil quality, texture, erosion, landform, land change, and wildlife. It can go a step further along with artificial intelligence (AI) to monitor live food quality change (Mohammad et al., 2014; Wu et al., 2015; Siva balan, 2016).

5.9.1 Electronic NOSE or E-nose

Electronic NOSE or E-nose (natural olfactory sensor emulator) can detect hydrophobic volatile gases from food products having a molecular weight in the range of 30–300 Da (Da). These are an array of chemical sensors with partial specificity and appropriate "pattern-recognition system" capable of identifying simple and complex odors. *E*-nose is composed of a sample loading part, sensor part, and a data processing unit, all acting as a unit. The sample loading unit transfers the volatile sample to the sensor part, which results in the change in resistance of sensors proportional to the applied sample. Volatile gases with high pressure from the sample such as alcohol, fatty acids due to yeast or bacterial fermentation on adsorption initiate a redox reaction. The data processing unit through intelligent pattern recognition algorithms, principal component analysis (PCA), discriminant factorial analysis, and partial least square quantify bitterness from the unrecognized samples. Nanocarriers with a large surface-area-to-volume ratio facilitate gas adsorption (Shevade et al., 2003). *E*-nose is successfully utilized for the reliable, sensitive, selective, and rapid detection of *Ganoderma boninense* responsible for stem rot on tree trunks and food products (Durán and Marcato, 2013). *E*-nose utilizing polymeric nanoparticles of polyanilineboronic acid can detect ongoing spoilage on stored food grains, and differentiating species producing different mycotoxins (Neethirajan et al., 2013). E-nose equipped with nanocomposites of conducting polymers like poly(vinyl butyral) and poly(ethylene oxide) with carbon black nanotubes and intelligent recognition pattern can detect *E.coli, Salmonella, S. aureus* from sausages and beef samples present in trace amounts, viz. below 50 ppb (Abdallah et al., 2013) (Fig. 13).

Mass spectroscopy (MS-related) coupled with *E*-nose can differentiate *Lactobacillus casei* species and can also discriminate between the cheese produced from the milk of healthy cows and those produced from the milk of a cow with mastitis (Trihaas et al., 2005). *E*-nose equipped with an MS detector employing SPME sampling technique can classify honey according to their botanical origin, geographical origin, viz. acacia, dandelion, chestnut, grape, linden, and fir (Ampuero et al., 2004). It can also predict various physicochemical parameters of honey such as fructose, glucose, hydroxymethylfurfural (HMF), amylase activity, moisture, ash content, and acidity (Huang et al., 2015).

E-nose coupled with a matrix of MOS gas sensors was used to discriminate arabica coffee counterfeited with cheap robusta coffee (Brudzewski et al., 2012). *E*-nose with six metal oxide gas sensors was used to classify different species of robusta coffee (Dong et al., 2017). Similarly, *E*-nose with six MOS gas sensors was used to separate Brazilian coffee from Guatemalan, Ethiopian, and Costa Rican coffee samples (Marek et al., 2020). A constellation of *E*-nose (10 MOS sensors), E-tongue (potentiometric), and electronic eye was used to determine the catechin, caffeine, polyphenol, and amino acid content of tea samples (Xu et al., 2019).

E-nose with a MOS sensor array was used to determine concentrations of bacteria ($> 10^2$cfu/ml) *Alicyclobacillus acidoterrestris* and *Alicyclobacillus acidocaldarius* in peach, orange, and apple fruit juices (Gobbi et al., 2010). PEN2 *E*-nose with 10 different MOS sensors and a potentiometric Alpha-Astree

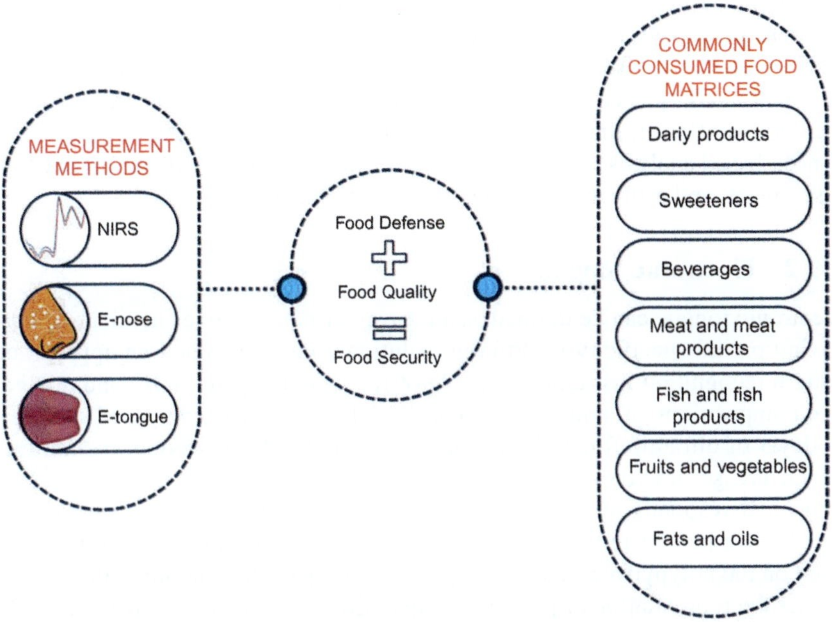

FIG. 13 Correlative analytical methods used as food quality control devices, their applications, and some of the commonly consumed food matrices defined by the Codex General Standard for Food Additives (Aouadi et al., 2020). *Copyright © 2020 by the authors. Licensee MDPI, Basel, Switzerland. This article is an open-access article distributed under the terms and conditions of the Creative Commons Attribution (CC BY) license (http://creativecommons.org/licenses/by/4.0/).*

E-tongue was used to detect fresh cherry tomato juices adulterated with over-ripe tomato juices (Hong and Wang, 2014). Ultra-fast GC-based Heracles II *E*-nose was used to detect the adulteration of apple juice with orange juice (Różańska et al., 2018). E-nose equipped with six thin-film MOS sensors was used to detect the contamination of *Alicyclobacillus* spp. in flavored drinks (Concina et al., 2010). Geosmin (GSM) and 2-methylisoborneol (MIB), two chief indicators of water contamination by the bacterial system were detected by *E*-nose with two vesicles (Lu and Teng, 2013).

E-nose with six MOS sensors was used to determine beef freshness stored for 6-days at −20°C; eight MOS sensors were used to determine beef decay. E-nose with 10 different MOS sensors was used to determine physiochemical parameters such as sensory scores, total volatile basic nitrogen, microbial population of beef, and adulteration of mutton with pork (Zhang et al., 2008; Hong et al., 2012; Tian et al., 2013).

E-nose with10 MOSFET sensors coupled with counter propagation artificial neural networks (CP-ANN) was used to determine the authenticity of extra virgin olive oils (EVOOs). In another study, E-nose with 10 MOSFET sensors and 12 MOS sensors in conjunction with E-tongue was used to determine the oxidation of EVOOs at different storage conditions. PEN2 E-nose with 10 MOS gas sensors was used to detect the adulteration of sesame oil with maize oil. PEN3 portable E-nose with 10 MOS sensors was applied to evaluate lipid oxidation in a variety of edible oils such as olive, peanut, soybean, rapeseed, camellia, corn, sunflower, linseed, and walnut (Triyana et al., 2015; Xu et al., 2016). E-nose-based analysis has promising application in fruit classification, ripeness assessment, and estimation of other quality parameters.

5.9.2 Electronic tongue

Electronic tongue can be defined as an "array of nonselective chemical sensors having partial specificity to different solution components and an appropriate pattern recognition instrument capable of recognizing quantitative and qualitative compositions of simple and complex solutions". It is composed of layer-by-layer of ultrathin films of molecular building lipids or polymer membranes which are several thousand-fold more sensitive (> 10,000) than the human tongue. It responds to the soluble portions of food through an array of sensors having pattern-recognition analysis. Receptors for bitterness are immobilized on the polypyrrole-modified carbon nanotubes. The binding of the analyte causes the reorientation of polypyrrole quantified as a signal (Hou et al., 2012). Utilizing electronic tongue, bitterness up to 1 fm can be detected from the beverages (Song et al., 2012). Degradation products in orange juices can be identified using electrodes of polyaniline-nanofibers working as taste sensors (Medeiros et al., 2009). *E*-tongues can be potentiometric, voltametric, and impedimetric. It usually works in absorbance, reflectance, and fluorescence mode (Aouadi et al., 2020) (Fig. 13).

Poly(ethylene glycol dimethacrylate-*N*methacryloyl-l-histidinemethylester) nanoparticles imprinted with chloramphenicol attached to SPR surface detected chloramphenicol from honey samples. It has also detected other related structures such as thiamphenicol and florfenicol (Kara et al., 2013). An impedimetric sensor was utilized to detect Penicillin G by immobilizing anti-penicillin G on gold electrodes functionalized with a self-assembled nanolayer of thioctic acid (Thavarungkul et al., 2007). A QCM nano-sensor with a gold surface containing a self-assembled monolayer (SAM) of mercaptans was used to determine tobramycin from chicken egg white and milk (Yola et al., 2014). An immunodipstick colloidal gold-coated with B12–IgY Ab was used to determine vitamin B12 measurements in food samples (Selva Kumar and Thakur, 2011). Carbon nanotubes (CNTs) coated with polymeric enzymes can detect cholesterol from poultry, meat, shrimp, and cheese samples employing voltammetry (Yang et al., 2011).

A potentiometric *E*-tongue was utilized to detect the adulteration of goat milk with bovine milk or urea-tampered milk (Dias et al., 2009; Wei and Wang, 2011). A voltammetric *E*-tongue was applied to evaluate bovine milk spiked with four different concentrations of antibiotics (Wei and Wang, 2011). Voltammetric

E-tongue was also applied to monitor yogurt samples during fermentation, ripening, and storage (Wei et al., 2017). Voltammetric E-tongue was used to analyze glucose, inverted sugar, and inulin syrups adulterated with honey samples. It can be used to determine the botanical origin of acacia, linden, sunflower, honeydew, and multifloral honey to decipher the geographical origin of multifloral honey (Oroian et al., 2018; Ciursa and Oroian, 2021; Sobrino-Gregorio et al., 2020). Potentiometric E-tongue was deployed to evaluate catechin, amino acid, polyphenol, and caffeine content in six different grades of tea (Xu et al., 2019). A piezoelectric E-tongue was employed to analyze the quinine content of tonic water. An E-tongue equipped with a lipidic and polymeric membrane was used for the estimation of glucose and fructose in soft drinks (Dias et al., 2011). A screen-printed electrode (SPE) modified voltammetric E-tongue based on bisphthalocyanine and polypyrrole was used to detect ammonia and putrescine as markers of the freshness of beef with high accuracy (Apetrei and Apetrei, 2016).

5.10. Food packaging

Efficient food packaging ensures a long shelf-life, sustainable safety, and quality of food. Nanotechnology can assist food packaging technology by aligning with antimicrobial and antifouling, food protection from moisture and oxygen, spoilage detection, and monitoring storage conditions. Nanotechnology application in food packaging can significantly reduce the events of food-borne diseases and hence lowering food wastage, food deterioration, and spoilage. Titanium dioxide, being a food additive and nontoxic can be applied for food packaging purposes. However, its in vivo physiological effect post-digestion needs to be thoroughly investigated (Dudefoi et al., 2018). Nanofibers made of cellulose can be explored for food packaging purposes owing to their biodegradability and bio-renewability (Abdul Khalil et al., 2016). Nanodiamonds being anti-inflammatory and antibacterial can be explored for food packaging. Nano-diamonds improve the flexibility, durability, resistance to humidity, and temperature of the packaged food material (Mitura et al., 2017).

5.10.1 Nanocoating

Nanocoating is the most widespread nano-technique for improving food packaging properties. Nano thin layers and nanofilms can be obtained from edible or nonedible food materials. Those obtained from edible food material can be applied directly while those obtained from nonedible material offer better protection (Joye et al., 2016). The edible nanocoatings can be applied via spraying, immersion, and rubbing (Zambrano-Zaragoza et al., 2018). These can be incorporated with antimicrobial and antioxidants to increase the quality of packaged food material (Wu et al., 2012). The nanofilms and nano thin layer cover the food to provide a mass transfer barrier. The edible nano-coatings can be used in the meat, agriculture industry, cheese, and bakery industry imparting color, flavor, enzyme, anti-oxidants, and anti-browning compounds (Nile et al., 2020). Other nanomaterials such as nano-clay are used in the packaging industry for their thermal, mechanical, economical, and barrier properties; while natural polyelectrolytes (protein, polysaccharides), charged lipids (phospholipids, surfactants), and colloidal particles (micelles, vesicles, droplets) can be utilized as nanolaminates in the food packaging industry (Gabr et al., 2015; Dasgupta et al., 2016). Nanolaminates can also be integrated with additional compounds such as antimicrobials, antioxidants, anti-browning agents, enzymes, flavors, odor, etc. to enhance the quality of packaged food material (Galus et al., 2020). Clay/polymer nanocomposites also improve the properties of the packaged food materials by imparting mechanical and barrier properties and enhanced shelf life, are lightweight, heat-resistant, and shatterproof. Current research is focused on the bio-composites integrated with natural fibers (Primož et al., 2021).

5.10.2 Smart and active packaging technologies

Smart and active packaging technologies preserve the packaged food items and informs the consumers about the spoilage. It involves the use of absorbing or emitting sachets/pads such as absorbent sachets containing O_2 scavengers; ethylene scavengers to reduce the fruit and vegetable ripening; decrease fat oxidation; humidity absorbers, odor absorbers, antimicrobial growth inhibitors; CO_2 emitting pads for inhibiting the microbial growth in packaged meat; preservative and antimicrobial releasers to extend the shelf-life of packaged food; and antioxidant releaser to prevent fat and oil oxidation (for references Mustafa and Andreescu, 2018). The antimicrobials act by either inhibiting the metabolic pathways or by degrading the cell walls/membranes. The antimicrobials can be natural extracts, fungicides (benomyl, imazalil), polymers (chitosan), bacteriocins (nisin, lauric acid), enzymes (lysozymes), organic acids (benzoic acid, sorbates), and nanoparticles (Ag, ZnO, TiO_2, MgO) (Primož et al., 2021). TiO_2 nanoparticles are nontoxic for human consumption and hence are approved the food packaging. Results have shown that a film containing Ag and TiO_2 shows the maximum antibacterial effect on gram-positive (*S. aureus*) and gram-negative bacteria (*E. coli*) (Lotfi et al., 2019; Chaudhary et al., 2020).

5.10.3 Nanoencapsulation

Nanoencapsulation during food processing and packaging entraps the essential oils and hence preserves and conserves their loss. Oil-loaded biopolymeric nanocarriers show enhanced antioxidant and antimicrobial properties. Plasticized polylactide (PLA) conjugated with bimetallic Ag-Cu nanoparticles and cinnamon essential oil is used for the packaging of chicken meat (Valdes et al., 2015). When it was tested against *Salmonella Typhimurium*, *Campylobacter jejuni*, and *Listeria monocytogenes*, it showed maximum antibacterial action for over 21 days stored at low temperature (Ahmed et al., 2018).

Similarly, buckwheat starch with ZnO nanoparticles showed good antibacterial activity against *Listeria monocytogenes* for mushroom packaging (Kim and Song, 2018). Montmorillonite clay and ginger extract with Ag nanoparticles used to synthesize antibacterial polyvinyl alcohol-based nanocomposite showed enhanced antibacterial activity against *S. Typhimurium* (Gram-negative bacteria) and *S. aureus* (Gram-positive bacteria) (Mathew et al., 2019). Polylactide/poly(Ɛ-caprolactone)/ZnO/clove essential oil (PLA/PEG/PCL/ZnO/CEO) composite antimicrobial films were utilized for the packaging scrambled eggs. It showed antibacterial activity against *S. aureus* and *E. coli* for 21 days (Ahmed et al., 2019). A biocomposite film having tragacanth/hydroxypropyl methylcellulose/beeswax with AgNPs showed antibacterial activity against the gram-positive pathogen (*B. cereus*, *S. aureus*, *Streptococcus pneumoniae*, and L. *monocytogenes*) and gram-negative (*E. coli*, *S. typhimorum*, *P. aeruginosa*, and *Klebsiella pneumoniae*) in a dose-dependent manner (Bahrami et al., 2019).

5.10.4 Enzymes

Enzymes with the potent antimicrobial and antioxidant properties are promising for active food packaging. However, reduction of protein stability in solution hampers their use. Lysozyme conjugated with nanocellulose showed enhanced antifungal and antibacterial activity against *C. albicans*, *A. niger*, *S. aureus*, and *E. coli* (Jebali et al., 2013). Lysozyme with lactoferrin incorporated on carboxymethyl cellulose was used for packaging raw meat and showed a good antimicrobial effect against *Listeria* (Barbiroli et al., 2012). β-Galactosidase from *Aspergillus oryzae* electrospunned into polyethylene oxide (PEO) nanofibers with polyethylene oxide/polypropylene oxide block copolymer facilitated dry storage (Wong et al., 2014).

Sensors with pH indicators are designed for showing the real-time spoilage of food items due to change in pH on account of the generation of volatile amines due to spoilage of packaged meat and fish. Several colorimetric enzymatic biosensors are designed to quantify the level of hypoxanthine in packaged fish, which is the main freshness indicator in fish (Lawal and Adeloju, 2012). An electrochemical sensor having immobilized xanthine oxidase on carbon paste electrode (CPE) functionalized with AuNPs was utilized for the detection of spoilage in meat and chicken (Agüí et al., 2006). A colorimetric sensor for the detection of xanthine was fabricated using copper nanoclusters and in another gold nanorods. Trimethylamine oxide indicates the level of fish freshness. It degrades to trimethyl oxide responsible for fishy odor. Trimethyl oxide can be detected by a colorimetric sensor using a pH indicator dye immobilized on cellulose microparticles (Huang et al., 2016; Chen et al., 2017a, b; Yan et al., 2017).

CO_2 emission due to the infestation from insects leading to mycotoxin production is the main indicator of food grain spoilage. A CO_2 sensor based on polyaniline boronic acid (PABA) was used to detect CO_2 in the simulated bulk of food grains (Neethirajan et al., 2010). An impedimetric aptasensor was designed to detect gluten in food samples. A sensor-based on a floating-gate transistor with antibody and aptamer as receptor was used for the detection of gluten from wheat and barley (White et al., 2018).

Intelligent packaging is coupled with sensing indicators or, radio frequency identification (RFID) tags connected to sensors. The sensing indicators provide real-time information about the qualitative and visual status of the packaged foods through a color code indicator or a dye diffusion. The thermochromic indicators show color change with the change of temperature. The thermochromic indicators contain thermochromic inks, which can be touch activated, cold activated, and high temperature-activated. A chemo-resistive ethylene sensor with single-walled carbon nanotubes (SWNTs) mixed with copper on a gold electrode was designed for measuring the level of ripening in fruits and vegetables (Esser et al., 2012).

Nano-silver with antibacterial agents can inhibit the growth of multiple bacteria. TiO_2 can enhance the UV protection function of the packaged material while simultaneously inhibiting bacterial growth (Ranjan et al., 2014). A nanocomposite of nylon and nano-clay can selectively keep the oxygen out and carbonation in for polyethylene bottles. Beer, soda, and milk can be kept for longer durations in nanocomposite bottles (Gruère, 2012). Adding enzymes such as cholesterol reductase or lactase enhances the nutritional value of food (Ranjan et al., 2014). Nano-barcodes equipped with synthetic DNA barcodes can detect the pathogen by UV fluorescence (Brody et al., 2008).

5.10.5 Bionanocomposites for smart food packaging

Starch and its derivatives such as PLA, polyhydroxy butyrate (PHB), polybutylene succinate (PBS), and aliphatic polyester, chitosan, proteins, and cellulose are preferably used as bio-nanocomposites for smart food packaging. Starch when combined with other nanomaterials such as TiO_2, ZnO nanoparticles, graphene, and poly(methyl methacrylate-*co*-acrylamide) can lead to improved mechanical, UV, and water-barrier properties. Graphene as an adduct can improve the mechanical properties, transmittance, and water vapor permeability of the composite film (starch/PVA/graphene oxide) (Wu et al., 2019a, b). These can be used for biodegradable food packaging.

The hybrid nanomaterials incorporating cellulose such as cellulose nanofibrils, cellulose nanocrystals, bacterial nanocellulose, and other hybrid nanomaterials can be used for food packaging. Various features of cellulose such as nanoscale structure, and high surface area imparts mechanical, biodegradation, optical, and barrier properties to cellulose nanocomposites. Packaging films from bacterial cellulose showed improved UV-absorbance and antioxidant properties (Sa et al., 2020).

Chitosan is inexpensive and commercially available. However, it comes up with poor thermal, mechanical, and barrier properties. The addition of nanomaterials can improve these properties. Low molecular weight chitosan shows improved antimicrobial properties. The antimicrobial character of chitosan owes it to binding with negatively charged bacterial cell-wall causing its disruption. Chitosan also acts as a chelating agent, binding to trace metals leading to toxin production and subsequent arrest of microbial growth (Yilmaz, 2020). Ag nanoparticles/chitosan/PVA fibrous composite nanolayers were used for packaging meat with extended shelf life and bioactivity (Pandey et al., 2020). ZnO NPs/linseed oil in chitosan/potato protein-based polymer was used for determining the storage quality of raw meat (Wang et al., 2020) (Fig. 14).

Robotic technology can impart anti-counterfeiting and anti-contaminant sensing ability to the packaged food material (Dilmaçünal, 2017). The biggest drawback of the intelligent food packaging technologies involving robotics, and nanotechnology is the cost of the packaging, which may shoot up to 100% of the packaged food material, however, ideally and by the acceptable norms, it must not exceed 10% of the packaged food material. Research is underway for

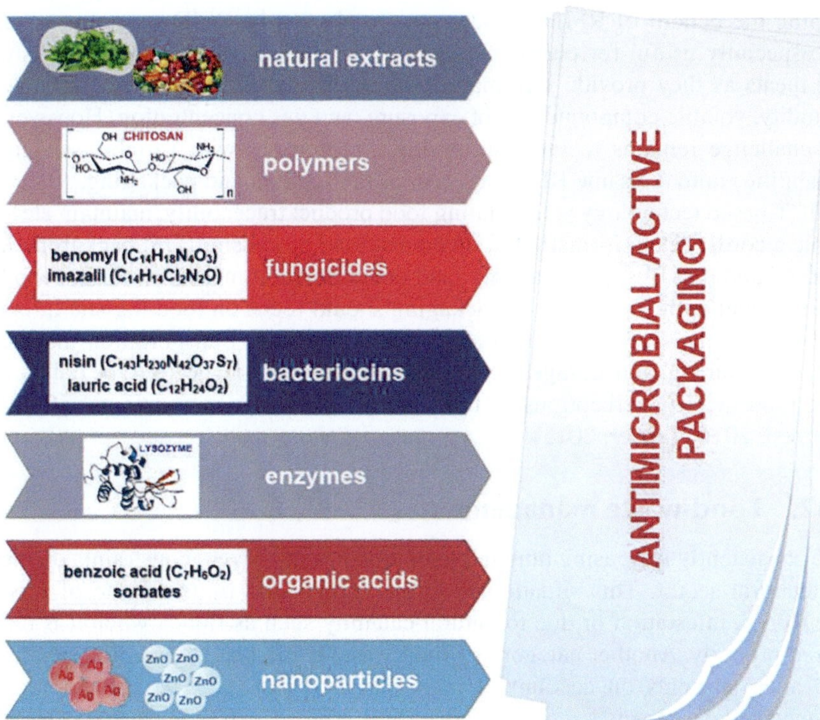

FIG. 14 Different antimicrobial substances for antimicrobial active packaging application (Primož et al., 2021). *Copyright © 2021 by the authors. Licensee MDPI, Basel, Switzerland. This article is an open-access article distributed under the terms and conditions of the Creative Commons Attribution (CC BY) license (http://creativecommons.org/licenses/by/4.0/).*

developing and improvising "Release command" packaging with a nanobioswitch that will release the preservatives when the food is about to spoil.

5.11. Supply chain

The food supply chain disseminates information regarding food attributes, animal welfare, GM issues, and country of origin to the customer. In this regard, counterfeit integrated circuit technology (CICT), radio frequency identification (RFID) technology, and information and communication technology (ICT) manage and regulates food traceability. RFID is a device having a microchip and antenna carrying information about the product. RFID is functionally analogous to barcodes, magnetic inks, biometrics, and voice recognition. Based on data transfer distances and power sources, RFID can be of three types, passive, semipassive and active tags. Supply chain giants from the USA such as Walmart and BestBuy have already employed the RFID tag and are promoting the same. Sensors integrated with RFID tags provide rapid, and secure information efficiently with consumer satisfaction. However, multinational food suppliers are reaping the benefit of RFID (Vanderroost et al., 2014). RFID tagged sensors are especially useful for perishable food items such as vegetables, fruit, fish, and meats as they provide information on temperature, pressure, pH, relative humidity, volatile compounds, light exposure, and gas concentration. However, the challenge remains to integrate multiple biosensors with variable sensing capabilities onto the same RFID tag, cost of fabrication, and packaging.

ICT-based technology is facilitating food product traceability, maintain electronic records, data information, food inputs, food processing, GM background, food-related diseases, pest tracking, and environmental measures. Future food management with respect to the packaging should focus on food logistic information such as packaging solutions, container loading, transportation modes, delivery planning, and storage conditions. This information needs to be collated with food quality, perceptions of food safety, and food environmental effects (Durresi, 2016; Tanner, 2012).

5.12. Food-waste management

The consistently increasing human population has put severe constraints on the agricultural sector. This situation is further aggravated by the waste of food due to pest infestation or due to natural calamity such as floods when it is not stored properly. Another category of food waste is the loss of essential macro- and micronutrients on account of improper food storage. The collaboration of noncovalent enzymes, clay minerals, biosensing electrodes can reduce the loss of moisture from the stored food. Nanomaterials like gold nanoparticles (AuNPs), quantum dots (QDs), carbon nanotubes (CNTs), nanowires can monitor and control the micro-environment of food. Nanotechnology in conjugation with photo-purification techniques can reliably assess, manage, and control milk products thus effectively replacing milk pasteurization with environment-friendly green solutions (Neethirajan et al., 2018a, b).

5.13. Biosensing in food engineering

Soil microbes are a crucial component of the soil ecosystem with a significant impact on food sustainability. "Soil ecological engineering" can modulate soil microbiota affecting soil diversity, enhance effective land use, reduce yield gap, and food diversity ensuring food availability in an environment-friendly and sustainable manner.

Metabolic engineering can be an important alternative to produce engineered energy, chemicals, and other rare materials. It can be further applied to design a sensor for the detection of flavonoids, or flavonoid generating bacterial strains in the future (Ahmed et al., 2016; Chae et al., 2017).

Nanomaterial with activated charcoal can enhance antimicrobial properties; food-grade nanoemulsion can be used in fruit juices; integrated nanomicrobials can be used as a water-purifier; nanomaterial induced nutraceutical delivery; and plant extracts can be used for food packaging. Cellulose nanofibers are already used for similar purposes (Dasgupta et al., 2015). Precision agriculture-mediated extraction of flowers, leaves, roots, and stems from diverse species can be conjugated with nanomaterials to synthesize nanoparticles directly from the natural extracts.

Nanomaterial-directed green synthesis can reduce phenolic compounds, alkaloids, and terpenoids to nanoparticles. The nanoparticles will serve as "magic bullets" leading to enhance plant growth, site-specific delivery of nutrients, and resistance to pathogens (Duhan et al., 2017).

5.14. Biofuel

A significant percentage of the population uses nonrenewable resources as fuel for domestic usage. Over-dependence on the already dwindled fossil fuel strains the existing resources. There is an urgent need to have an alternative as a biofuel to meet the energy requirement for domestic purposes. Innovations, such as toggle switches, synthetic entities mimicking natural molecules, and gene networks facilitate the redesign of switchable functions of sensing elements. The sensors based on these such as biomimetics, molecular imprinting polymers (MIP), aptamers, peptide nucleic acids (PNAs), and ribozymes are successfully utilized for the production of biofuels (Scognamiglio et al., 2015).

5.15. Conclusion and future perspectives

The renewed interest in food quality and safety requires the development of methods that are sensitive and reliable for the detection of food contaminants with high analytical accuracy in the background of complex food matrices ensuring freshness preservation. Conventional analytical techniques for food-related investigations are available, but it suffers from the inherent drawbacks of complicated analytical procedures, multi-step analytical process, the requirement of expensive infrastructural set-up, trained and skilled manpower, and lack of portability, etc. In this context, electrochemical biosensors have received tremendous attention owing to their operational simplicity, sensitivity, selectivity, specificity, stability, miniaturization ability, reproducibility, and low-cost analysis. Nanomaterials aided the advance design of biosensors for detecting and quantifying freshness markers, allergens, adulterants, pathogens, toxins, and spurious chemicals are already in place. Biosensing in the food industry facilitates smart packaging, food-status monitoring, shelf-life extension, food-waste reduction, and overall food quality improvement. Nanomaterials and nanocomposites in food packaging improve the mechanical strength, impart antimicrobial activity, and light-blocking properties. However, there are issues related to the migration of nanoparticles from packaging material to food. The tests to validate the proof of such migration and its potential impact on human health are still not clear. This presumed toxicity, lack of clinical data to validate the risk assessment limits the use of nanomaterials in food-related industries. The advanced chemometric tools such as electronic-nose (*E*-nose), electronic tongue (E-tongue), NIR-spectroscopy coupled with mathematical correction techniques are at the forefront of food-quality control by providing fingerprints of food properties, thereby detecting, classifying, and predicting food contaminants. However, biosensing techniques are still in their developmental stage and need validation of their application in real-world samples in diverse environmental setups. The issues of sensitivity, selectivity, robustness, and safety of the nanomaterials, and other components in biosensors due to potential contact or migration in food need to be established. An all-in-one-system biosensing platform is required that can facilitate the multifunctional properties for capture, separation, inactivation, and on-site detection of real-world samples in simple, rapid, accurate, sensitive, selective, specific, noninvasive, nondestructive, and the low-cost manner in diverse environmental setups. Key to the development of model biosensing platform with all these attributes requires targeted focus on the following five parameters: specific bioreceptors (enhancing sensitivity); microfluidics (automation); paper-based biosensors (detection and cost minimization), and smartphones or other mobile devices (simple signal readout).

References

Abdallah, S.A., Al-Shatti, L.A., Alhajraf, A.F., Al-Hammad, N., Al-Awadi, B., 2013. The detection of foodborne bacteria on beef: the application of the electronic nose. Springerplus 2, 687.

Abdul Khalil, H.P.S., Davoudpour, Y., Saurabh, C.K., Hossain, M.S., Adnan, A.S., Dungani, R., Paridah, M.T., Sarker, M.Z.I., Fazita, M.R.N., Syakir, M.I., Haafiz, M.K.M., 2016. A review on nanocellulosic fibres as new material for sustainable packaging: process and applications. Renew. Sustain. Energy Rev. 64, 823–836.

Adriano Aquino, A., Conte-Junior, C.A., 2020. A systematic review of food allergy: nanobiosensor and food allergen detection. Biosensors 10, 194.

Agüí, L., Manso, J., Yáñez-Sedeño, P., Pingarrón, J.M., 2006. Amperometric biosensor for hypoxanthine based on immobilized xanthine oxidase on nanocrystal gold-carbon paste electrodes. Sensors Actuators B Chem. 113, 272–280.

Ahmed, J., Arfat, Y.A., Bher, A., Mulla, M., Jacob, H., Auras, R., 2018. Active chicken meat packaging based on polylactide films and bimetallic Ag-Cu nanoparticles and essential oil. J. Food Sci. 83, 1299–1310.

Ahmed, S., Bui, M.-P.N., Abbas, A., 2016. Paper-based chemical and biological sensors: engineering aspects. Biosens. Bioelectron. 77, 249–263.

Ahmed, J., Mulla, M., Jacob, H., Luciano, G., Bini, T.B., Almusallam, A., 2019. Polylactide/poly(ε-caprolactone)/zinc oxide/clove essential oil composite antimicrobial films for scrambled egg packaging. Food Packag. Shelf Life 21, 100355.

Akram, M., Stuart, M.C., Wong, D.K.Y., 2004. Direct application strategy to immobilise a thioctic acid self-assembled monolayer on a gold electrode. Anal. Chim. Acta 504, 243–251.

Alamprese, C., Casale, M., Sinelli, N., Lanteri, S., Casiraghi, E., 2013. Detection of minced beef adulteration with Turkey meat by UV e vis, NIR and MIR spectroscopy. LWT Food Sci. Technol. 53, 225–232.

Alarcon-Angeles, G., Álvarez-Romero, G.A., Merkoçi, A., 2016. Emerging nanomaterials for analytical detection. In: Comprehensive Analytical Chemistry. vol. 74. Elsevier, Amsterdam, The Netherlands, pp. 195–246. ISBN 978-0-444-63579-2.

Alhogail, S., Suaifan, G.A., Zourob, M., 2016. Rapid colorimetric sensing platform for the detection of *Listeria monocytogenes* foodborne pathogen. Biosens. Bioelectron. 86, 1061–1066.

Ali, A.A., Altemimi, A.B., Alhelfi, N., Ibrahim, S.A., 2020. Application of biosensors for detection of pathogenic food bacteria: a review. Biosensors 10, 58.

Alves, R.C., Barroso, M.F., González-García, M.B., Oliveira, M.B.P.P., Delerue-Matos, C., 2016. New trends in food allergens detection: toward biosensing strategies. Crit. Rev. Food Sci. Nutr. 56, 2304–2319.

Amiri, M., Bezaatpour, A., Jafari, H., Boukherroub, R., Szunerits, S., 2018. Electrochemical methodologies for the detection of pathogens. ACS Sens. 3, 1069–1086.

Ampuero, S., Bogdanov, S., Bosset, J.O., 2004. Classification of unifloral honeys with an MS-based electronic nose using different sampling modes: SHS, SPME and INDEX. Eur. Food Res. Technol. 218, 198–207.

Aouadi, B., Zaukuu, J.-L.Z., Vitális, F., Bodor, Z., Fehér, O., Gillay, Z., Bazar, G., Kovacs, Z., 2020. Historical evolution and food control achievements of near infrared spectroscopy, electronic nose, and electronic tongue-critical overview. Sensors (Basel) 20, 5479.

Apetrei, I.M., Apetrei, C., 2016. Application of voltammetric e-tongue for the detection of ammonia and putrescine in beef products. Sensors Actuators B Chem. 234, 371–379.

Arduini, F., Cinti, S., Scognamiglio, V., Moscone, D., 2016. Nanomaterials in electrochemical biosensors for pesticide detection: advances and challenges in food analysis. Microchim. Acta 183, 2063–2083.

Arlett, J.L., Myers, E.B., Roukes, M.L., 2011. Comparative advantages of mechanical biosensors. Nat. Nanotechnol. 6, 203.

Artigues, M., Abellà, J., Colominas, S., 2017. Analytical parameters of an amperometric glucose biosensor for fast analysis in food samples. Sensors 17, 2620.

Bahadır, E.B., Sezgintürk, M.K., 2017. Biosensor technologies for analyses of food contaminants. In: Nanobiosensors. Elsevier, Amsterdam, The Netherlands, pp. 289–337. ISBN 978-0-12-804301-1.

Bahrami, A., Rezaei Mokarram, R., Sowti Khiabani, M., Ghanbarzadeh, B., Salehi, R., 2019. Physico-mechanical and antimicrobial properties of tragacanth/hydroxypropyl methylcellulose/ beeswax edible films reinforced with silver nanoparticles. Int. J. Biol. Macromol. 129, 1103–1112.

Bakke, M., Suzuki, S., 2018. Development of a novel hygiene monitoring system based on the detection of total adenylate (ATP+ ADP+ AMP). J. Food Prot. 81, 729–737.

Barbiroli, A., Bonomi, F., Capretti, G., Iametti, S., Manzoni, M., Piergiovanni, L., Rollini, M., 2012. Antimicrobial activity of lysozyme and lactoferrin incorporated in cellulose-based food packaging. Food Control 26, 387–392.

Barbosa, J., Albano, H., Silva, C.P., Teixeira, P., 2019. Microbiological contamination of reusable plastic bags for food transportation. Food Control 99, 158–163.

Batra, B., Lata, S., Sharma, M., Pundir, C.S., 2013. An acrylamide biosensor based on immobilization of hemoglobin onto multiwalled carbon nanotube/copper nanoparticles/polyaniline hybrid film. Anal. Biochem. 433, 210–217.

Bázár, G., Romvári, R., Szabó, A., Somogyi, T., Éles, V., Tsenkova, R., 2016. NIR detection of honey adulteration reveals differences in water spectral pattern. Food Chem. 194, 873–880.

Bazar, G., Szabo, A., Andrassy-Baka, G., Romvari, L., Locsmandi, R., Princz, Z., Jekkel, G., 2007. NIRS prediction for protein and intramuscular fat content of rabbit hind leg meat. Agriculture 13, 1–5.

Bertone, E., Venturello, A., Giraudo, A., Pellegrino, G., Geobaldo, F., 2016. Simultaneous determination by NIR spectroscopy of the roasting degree and Arabica/Robusta ratio in roasted and ground coffee. Food Control 59, 683–689.

Brody, A.L., Bugusu, B., Han, J.H., Sand, C.K., McHugh, T.H., 2008. Scientific status summary: innovative food packaging solutions. J. Food Sci. 73, R107–R116.

Brudzewski, K., Osowski, S., Member, S., Dwulit, A., 2012. Recognition of coffee using differential electronic nose. IEEE Trans. Instrum. Meas. 61, 1803–1810.

Bunney, J., Williamson, S., Atkin, D., Jeanneret, M., Cozzolino, D., Chapman, J., Power, A., Chandra, S., 2017. The use of electrochemical biosensors in food analysis. Curr. Res. Nutr. Food Sci. 5, 183–195.

Bwambok, D.K., Siraj, N., Macchi, S., Larm, N.E., Baker, G.A., Pérez, R.L., Ayala, C.E., Walgama, C., Pollard, D., Rodriguez, J.D., Banerjee, S., Elzey, B., Warner, I.M., Fakayode, S.O., 2020. QCM sensors arrays, electroanalytical techniques and NIR spectroscopy coupled to mulativariate analysis for quality assessment of food products, raw materials, ingredients and foodborne pathogen detection: challenges and breakthrough. Sensors (Basel) 20, 6982.

Callejas, L., Darce, A.C.M., Amador, J.J., Conklin, L., Gaffga, N., Schurz Rogers, H., DeGrasse, S., Hall, S., Earley, M., Mei, J., et al., 2015. Paralytic shellfish poisonings resulting from an algal bloom in Nicaragua. BMC. Res. Notes 8, 74.

Campos, M.I., Mussons, M.L., Antolin, G., Debán, L., Pardo, R., 2017. On-line prediction of sodium content in vacuum packed dry-cured ham slices by non-invasive near infrared spectroscopy. Meat Sci. 126, 29–35.

Campuzano, S., Montiel, V.R., Serafín, V., Yáñez-Sedeño, P., Pingarrón, J.M., 2020. Cutting-edge advances in electrochemical affinity biosensing at different molecular level of emerging food allergens and adulterants. Biosensors 10, 10.

Campuzano, S., Yáñez-Sedeño, P., Pingarrón, J.M., 2019. Nanoparticles for nucleic-acid-based biosensing: opportunities, challenges, and prospects. Anal. Bioanal. Chem. 411, 1791–1806.

Castell-Perez, E., Gomes, C., Tahtouh, J., Moreira, R., McLamore, E.S., Knowles, H.S., 2017. Food processing and waste within the nexus framework. Curr. Sustain. Energy Rep. 4, 99–108.

Chae, T.U., Choi, S.Y., Kim, J.W., Ko, Y.-S., Lee, S.Y., 2017. Recent advances in systems metabolic engineering tools and strategies. Curr. Opin. Biotechnol. 47, 67–82.

Chan, C., Guo, J., Sun, C., Tsang, M., Tian, F., Hao, J., Chen, S., Yang, M., 2015. A reduced graphene oxide-Au based electrochemical biosensor for ultrasensitive detection of enzymatic activity of botulinum neurotoxin A. Sensors Actuators B Chem. 220, 131–137.

Chaudhary, P., Fatima, F., Kumar, A., 2020. Relevance of nanomaterials in food packaging and its advanced future prospects. J. Inorg. Organomet. Polym. 30, 5180–5192.

Chen, J., 2015. Bacteriophage-based nanoprobes for rapid bacteria separation. Nanoscale 7, 16230.

Chen, J., Alcaine, S.D., Jiang, Z., Rotello, V.M., Nugen, S.R., 2015. Detection of *Escherichia coli* in drinking water using T7 bacteriophage-conjugated magnetic probe. Anal. Chem. 87, 8977–8984.

Chen, M., Gan, N., Li, T., Wang, Y., Xu, Q., Chen, Y., 2017a. An electrochemical aptasensor for multiplex antibiotics detection using y-shaped DNA-based metal ions encoded probes with NMOF substrate and CSRP target-triggered amplification strategy. Anal. Chim. Acta 968, 30–39.

Chen, K., Han, H., Luo, Z., Wang, Y., Wang, X., 2012. A practicable detection system for genetically modified rice by SERS-barcoded nanosensors. Biosens. Bioelectron. 34, 118–124.

Chen, Z., Lin, Y., Ma, X., Guo, L., Qiu, B., Chen, G., Lin, Z., 2017b. Multicolor biosensor for fish freshness assessment with the naked eye. Sensors Actuators B Chem. 252, 201–208.

Ciursa, P., Oroian, M., 2021. Voltametric E-tongue for honey adulteration detection. Sensors 21, 5059.

Concina, I., Bornšek, M., Baccelliere, S., Falasconi, M., Gobbi, E., Sberveglieri, G., 2010. *Alicyclobacillus* spp.: detection in soft drinks by electronic nose. Food Res. Int. 43, 2108–2114.

Craig, A.P., Botelho, B.G., Oliveira, L.S., Franca, A.S., 2018. Mid infrared spectroscopy and chemometrics as tools for the classification of roasted coffees by cup quality. Food Chem. 245, 1052–1061.

Danesh, N.M., Ramezani, M., Emrani, A.S., Abnous, K., Taghdisi, S.M., 2016. A novel electrochemical aptasensor based on arch-shape structure of aptamer-complimentary strand conjugate and exonuclease i for sensitive detection of streptomycin. Biosens. Bioelectron. 75, 123–128.

Dasgupta, N., Ranjan, S., Mundekkad, D., Ramalingam, C., Shanker, R., Kumar, A., 2015. Nanotechnology in agro-food: from field to plate. Food Res. Int. 69, 381–400.

Dasgupta, N., Ranjan, S., Patra, D., Srivastava, P., Kumar, A., Ramalingam, C., 2016. Bovine serum albumin interacts with silver nanoparticles with a "side-on" or "end on" conformation. Chem. Biol. Interact. 253, 100–111.

De Carvalho, B.M.A., De Carvalho, L.M., Dos Reis Coimbra, J.S., Minim, L.A., De Souza Barcellos, E., Da Silva Júnior, W.F., Detmann, E., De Carvalho, G.G.P., 2015. Rapid detection of whey in milk powder samples by spectrophotometric and multivariate calibration. Food Chem. 174, 1–7.

De Marchi, M., Manuelian, C.L., Manfrin, D., Meneghesso, M., Cassandro, M., Penasa, M., 2017. Prediction of sodium content in commercial processed meat products using near infrared spectroscopy. Meat Sci. 125, 61–65.

De Santana, M.C., Ferreira, M.M.C., Pallone, J.A.L., 2020. Control of ascorbic acid in fortified powdered soft drinks using near-infrared spectroscopy (NIRS) and multivariate analysis. J. Food Sci. Technol. 57, 1233–1241.

Deng, P., Xu, Z., Kuang, Y., 2014. Electrochemical determination of bisphenol a in plastic bottled drinking water and canned beverages using a molecularly imprinted chitosan-graphene composite film modified electrode. Food Chem. 157, 490–497.

Derda, R., Lockett, M.R., Tang, S.K., Fuller, R.C., Maxwell, E.J., Breiten, B., Cuddemi, C.A., Ozdogan, A., Whitesides, G.M., 2013. Filter-based assay for *Escherichia coli* in aqueous samples using bacteriophage-based amplification. Anal. Chem. 85, 7213–7220.

Dias, L.G., Peres, A.M., Barcelos, T.P., Sá Morais, J., Machado, A.A.S.C., 2011. Semi-quantitative and quantitative analysis of soft drinks using an electronic tongue. Sensors Actuators B Chem. 154, 111–118.

Dias, L.A., Peres, A.M., Veloso, A.C., Reis, F.S., Vilas-Boas, M., Machado, A.A., 2009. An electronic tongue taste evaluation: identification of goat milk adulteration with bovine milk. Sensors Actuators B Chem. 136, 209–217.

Dilmaçünal, T., 2017. Intelligent systems in the food packaging industry. In: Nanotechnology Applications in Food. Elsevier, Amsterdam, The Netherlands, pp. 287–306. ISBN 978-0-12-811942-6.

Doeun, D., Davaatseren, M., Chung, M.S., 2017. Biogenic amines in foods. Food Sci. Biotechnol. 26, 1463–1474.

Dong, W., Zhao, J., Hu, R., Dong, Y., Tan, L., 2017. Differentiation of Chinese robusta coffees according to species, using a combined electronic nose and tongue, with the aid of chemometrics. Food Chem. 229, 743–751.

Dudefoi, W., Villares, A., Peyron, S., Moreau, C., Ropers, M.-H., Gontard, N., Cathala, B., 2018. Nanoscience and nanotechnologies for biobased materials, packaging and food applications: new opportunities and concerns. Innov. Food Sci. Emerg. Technol. 46, 107–121.

Duhan, J.S., Kumar, R., Kumar, N., Kaur, P., Nehra, K., Duhan, S., 2017. Nanotechnology: the new perspective in precision agriculture. Biotechnol. Rep. 15, 11–23.

Durán, N., Marcato, P.D., 2013. Nanobiotechnology perspectives: role of nanotechnology in the food industry: a review. Int. J. Food Sci. Technol. 48, 1127–1134.

Durresi, M., 2016. (Bio)sensor integration with ICT tools for supplying chain management and traceability in agriculture. In: Comprehensive Analytical Chemistry. vol. 74. Elsevier, Amsterdam, The Netherlands, pp. 389–413. ISBN 978-0-444-63579-2.

Eangoor, P., Indapurkar, A.S., Vakkalanka, M., Yeh, J.S., Knaack, J.S., 2017. Rapid and sensitive ELISA screening assay for several paralytic shellfish toxins in human urine. J. Anal. Toxicol. 41, 755–759.

Eissa, S., Chinnappan, R., Zourob, M., 2017. Advances in biosensor technologies for food allergen monitoring and diagnosis. In: Abdel Rahman, A.M. (Ed.), Food Allergy Methods of Detection and Clinical Studies. Taylor & Francis Group, Abington, UK.

Eissa, S., Zourob, M., 2012. A graphene-based electrochemical competitive immunosensor for the sensitive detection of okadaic acid in shellfish. Nanoscale 4, 7593–7599.

Eissa, S., Zourob, M., 2017. In vitro selection of DNA aptamers targeting β-lactoglobulin and their integration in graphene-based biosensor for the detection of milk allergen. Biosens. Bioelectron. 91, 169–174.

Escarpa, A., 2014. Lights and shadows on food microfluidics. Lab Chip 14, 3213–3224.

Esser, B., Schnorr, J.M., Swager, T.M., 2012. Selective detection of ethylene gas using carbon nanotube-based devices: utility in determination of fruit ripeness. Angew. Chem. Int. Ed. 51, 5752–5756.

Fan, E., Peng, J., Shi, Y., Ouyang, H., Xu, Z., Fu, Z., 2020. Quantification of live Gram-positive bacteria via employing artificial antibacterial peptide-coated magnetic spheres as isolation carriers. Microchem. J. 154, 104643.

Gabr, M.H., Okumura, W., Ueda, H., Kuriyama, W., Uzawa, K., Kimpara, I., 2015. Mechanical and thermal properties of carbon fiber/polypropylene composite filled with nano-clay. Compos. Part B 69, 94–100.

Gallo, M., Ferranti, P., 2016. The evolution of analytical chemistry methods in foodomics. J. Chromatogr. A 1428, 3–15.

Galus, S., Arik Kibar, E.A., Gniewosz, M., Kraśniewska, K., 2020. Novel materials in the preparation of edible films and coatings—a review. Coatings 10, 674.

Garcia, A.B., 2013. The use of data mining techniques to discover knowledge from animal and food data: examples related to the cattle industry. Trends Food Sci. Technol. 29, 151–157.

Gaudin, V., 2017. Advances in biosensor development for the screening of antibiotic residues in food products of animal origin—a comprehensive review. Biosens. Bioelectron. 90, 363–377.

Gobbi, E., Falasconi, M., Concina, I., Mantero, G., Bianchi, F., Mattarozzi, M., Musci, M., 2010. Electronic nose and *Alicyclobacillus* spp. spoilage of fruit juices: an emerging diagnostic tool. Food Control 21, 1374–1382.

Griffiths, M.W., 1996. The role of ATP bioluminescence in the food industry: new light on old problems. Food Technol. 50, 62–72.

Gruère, G.P., 2012. Implications of nanotechnology growth in food and agriculture in OECD countries. Food Policy 37, 191–198.

Haddada, M.B., Salmain, M., Boujday, S., 2018. Gold colloid-nanostructured surfaces for enhanced piezoelectric immunosensing of staphylococcal enterotoxin a. Sensors Actuators B Chem. 255, 1604–1613.

Hayat, A., Catanante, G., Marty, J.L., 2014. Current trends in nanomaterial-based amperometric biosensors. Sensors 14, 23439–23461.

Hiep, H.M., Endo, T., Kerman, K., Chikae, M., Kim, D.-K., Yamamura, S., Takamura, Y., Tamiya, E., 2007. A localized surface plasmon resonance based immunosensor for the detection of casein in milk. Sci. Technol. Adv. Mater. 8, 331–338.

Hong, X., Wang, J., 2014. Detection of adulteration in cherry tomato juices based on electronic nose and tongue: comparison of different data fusion approaches. J. Food Eng. 126, 89–97.

Hong, X., Wang, J., Hai, Z., 2012. Discrimination and prediction of multiple beef freshness indexes based on electronic nose. Sensors Actuators B Chem. 161, 381–389. Sensors and Actuators B: Chemical.

Hou, Y., Genua, M., Tada Batista, D., Calemczuk, R., Buhot, A., Fornarelli, P., Koubachi, J., Bonnaffe, D., Saesen, E., Laguri, C., Lortat-Jacob, H., Livache, T., 2012. Continuous evolution profiles for electronic-tongue-based analysis. Angew. Chem. Int. Ed. Eng. 51, 10394–10398.

Hu, T., Mei, X., Wang, Y., Weng, X., Liang, R., Wei, M., 2019. Two-dimensional nanomaterials: fascinating materials in biomedical field. Sci. Bull. 64, 1707–1727.

Huang, X., Li, Z., Zou, X., Shi, J., Mao, H., Zhao, J., Hao, L., Holmes, M., 2016. Detection of meat-bornetrimethylamine based on nanoporous colorimetric sensor arrays. Food Chem. 197, 930–936.

Huang, L., Liu, H., Zhang, B., Wu, D., 2015. Application of electronic nose with multivariate analysis and sensor selection for botanical origin identification and quality determination of honey. Food Bioprocess Technol. 8, 359–370.

Huang, F., Song, H., Guo, L., Guang, P., Yang, X., Li, L., Zhao, H., Yang, M., 2020. Detection of adulteration in Chinese honey using NIR and ATR-FTIR spectral data fusion. Spectrochim. Acta A Mol. Biomol. Spectrosc. 235, 118297.

Hughes, G., Pemberton, R.M., Fielden, P.R., Hart, J.P., 2016. The design, development and application of electrochemical glutamate biosensors. TrAC Trends Anal. Chem. 79, 106–113.

Ibrišimović, N., Ibrišimović, M., Kesić, A., Pittner, F., 2015. Microbial biosensor: a new trend in the detection of bacterial contamination. Chem. Mon. 146, 1363–1370.

Idili, A., Parolo, C., Ortega, G., Plaxco, K.W., 2019. Calibration-free measurement of phenylalanine levels in the blood using an electrochemical aptamer-based sensor suitable for point-of-care applications. ACS Sens. 4, 3227–3233.

Ilaslan, K., Boyaci, I.H., Topcu, A., 2015. Rapid analysis of glucose, fructose and sucrose contents of commercial soft drinks using Raman spectroscopy. Food Control 48, 56–61.

Istamboulié, G., Paniel, N., Zara, L., Granados, L.R., Barthelmebs, L., Noguer, T., 2016. Development of an impedimetric aptasensor for the determination of aflatoxin M1 in milk. Talanta 146, 464–469.

Jayan, H., Pu, H., Sun, D.W., 2019. Recent development in rapid detection techniques for microorganism activities in food matrices using bio-recognition: a review. Trends Food Sci. Technol. 95, 233–246.

Jebali, A., Hekmatimoghaddam, S., Behzadi, A., Rezapor, I., Mohammadi, B.H., Jasemizad, T., Yasini, S.A., Javadzadeh, M., Amiri, A., Soltani, M., et al., 2013. Antimicrobial activity of nanocellulose conjugated with allicin and lysozyme. Cellulose 20, 2897–2907.

Jiang, H., Jiang, D., Shao, J., Sun, X., 2016b. Magnetic molecularly imprinted polymer nanoparticles based electrochemical sensor for the measurement of gram-negative bacterial quorum signaling molecules (*N*-acyl-homoserine-lactones). Biosens. Bioelectron. 75, 411–419.

Jiang, H., Jiang, D., Zhu, P., Pi, F., Ji, J., Sun, C., Sun, J., Sun, X., 2016a. A novel mast cell co-culture microfluidic chip for the electrochemical evaluation of food allergen. Biosens. Bioelectron. 83, 126–133.

Jiang, Y., Yang, X., Liang, P., Liu, P., Huang, X., 2018. Microbial fuel cell sensors for water quality early warning systems: fundamentals, signal resolution, optimization and future challenges. Renew. Sust. Energ. Rev. 81, 292–305.

Joye, I.J., Davidov-Pardo, G., McClements, D.J., 2016. Nanotechnology in food processing. In: Caballero, B., Finglas, P.M., Toldrá, F. (Eds.), Encyclopedia of Food and Health. Academic Press, Oxford, UK, pp. 49–55. ISBN 978-0-12-384953-3.

Kamboj, U., Kaushal, N., Mishra, S., Munjal, N., 2020. Application of selective near infrared spectroscopy for qualitative and quantitative prediction of water adulteration in milk. Mater. Today Proc. 24, 2449–2456.

Kamruzzaman, M., Sun, D., Elmasry, G., Allen, P., 2013. Fast detection and visualization of minced lamb meat adulteration using NIR hyperspectral imaging and multivariate image analysis. Talanta 103, 130–136.

Kara, M., Uzun, L., Kolayli, S., Denizli, A., 2013. Combining molecular imprinted nanoparticles with surface plasmon resonance nanosensor for chloramphenicol detection in honey. J. Appl. Polym. Sci. 129, 2273–2279.

Kashyap, P.L., Xiang, X., Heiden, P., 2015. Chitosan nanoparticle based delivery systems for sustainable agriculture. Int. J. Biol. Macromol. 77, 36–51.

Kim, S., Song, K.B., 2018. Antimicrobial activity of buckwheat starch films containing zinc oxide nanoparticles against listeria monocytogenes on mushrooms. Int. J. Food Sci. Technol. 53, 1549–1557.

Kudr, J., Klejdus, B., Adam, V., Zitka, O., 2018. Magnetic solids in electrochemical analysis. TrAC Trends Anal. Chem. 98, 104–113.

Lawal, A.T., Adeloju, S.B., 2012. Progress and recent advances in fabrication and utilization of hypoxanthine biosensors for meat and fish quality assessment: a review. Talanta 100, 217–228.

Li, X., He, Y., Wu, C., Sun, D.W., 2007. Nondestructive measurement and fingerprint analysis of soluble solid content of tea soft drink based on Vis/NIR spectroscopy. J. Food Eng. 82, 316–323.

Li, H., Somerson, J., Xia, F., Plaxco, K.W., 2018. Electrochemical DNA-based sensors for molecular quality control: continuous, real-time melamine detection in flowing whole milk. Anal. Chem. 90, 10641–10645.

Liao, W., Lu, X., 2016. Determination of chemical hazards in foods using surface-enhanced Raman spectroscopy coupled with advanced separation techniques. Trends Food Sci. Technol. 54, 103–113.

Lin, H.-Y., Huang, C.-H., Park, J., Pathania, D., Castro, C.M., Fasano, A., Weissleder, R., Lee, H., 2017. Integrated magneto-chemical sensor for on-site food allergen detection. ACS Nano 1, 10062–10069.

Liu, X., Marrakchi, M., Xu, D., Dong, H., Andreescu, S., 2016a. Biosensors based on modularly designed synthetic peptides for recognition, detection and live/dead differentiation of pathogenic bacteria. Biosens. Bioelectron. 80, 9–16.

Liu, Y., Weiss, N.O., Duan, X., Cheng, H.-C., Huang, Y., Duan, X., 2016b. Van der Waals heterostructures and devices. Nat. Rev. Mater. 1, 16042.

Liu, B., Zhang, B., Cui, Y., Chen, H., Gao, Z., Tang, D., 2011. Multifunctional gold-silica nanostructures for ultrasensitive electrochemical immunoassay of streptomycin residues. ACS Appl. Mater. Interfaces 3, 4668–4676.

López, M.A., Moreno-Guzman, M., Jurado, B., Escarpa, A., 2016. Food microfluidics biosensors. In: Comprehensive Analytical Chemistry. Elsevier, Amsterdam, The Netherlands. ISBN 978-0-444-63579-2.

Lotfi, S., Ahari, H., Sahraeyan, R., 2019. The effect of silver nanocomposite packaging based on melt mixing and sol-gel methods on shelf life extension of fresh chicken stored at 4 °C. J. Food Saf. 39, e12625.

Lu, X., Teng, J., 2013. An improved methodology of soft drink discrimination using an electronic nose. In: Proceedings of the 2013 3rd International Conference on Computer Science and Network Technology, Dalian, China, 12–13 October, pp. 982–985.

Majdinasab, M., Yaqub, M., Rahim, A., Catanante, G., Hayat, A., Marty, J.L., 2017. An overview on recent progress in electrochemical biosensors for antimicrobial drug residues in animal-derived food. Sensors 17, 1947. 2017.

Malvano, F., Pilloton, R., Albanese, D., 2019. Label-free impedimetric biosensors for the control of food safety—a review. Int. J. Environ. Anal. Chem. 100, 1–24.

Mamani-linares, L.W., Gallo, C., Alomar, D., 2012. Identification of cattle, llama and horse meat by near infrared reflectance or transflectance spectroscopy. Meat Sci. 90, 378–385.

Mandal, P.K., Biswas, A.K., 2020. Modern techniques for rapid detection of meatborne pathogens. In: Meat Quality Analysis. Academic Press, Cambridge, MA, USA, pp. 287–303.

Marek, G., Dobrzański, B., Oniszczuk, T., Combrzyński, M., Cwikła, D., Rusinek, R., 2020. Detection and ′ differentiation of volatile compound profiles in roasted coffee arabica beans from different countries using an electronic nose and GC-MS. Sensors 20, 2124.

Marin, S., Ramos, A.J., Cano-Sancho, G., Sanchis, V., 2013. Mycotoxins: occurrence, toxicology, and exposure assessment. Food Chem. Toxicol. 60, 218–237.

Marquetti, I., Link, J.V., Lemes, A.L.G., dos Santos Scholz, M.B., Valderrama, P., Bona, E., 2016. Partial least square with discriminant analysis and near infrared spectroscopy for evaluation of geographic and genotypic origin of arabica coffee. Comput. Electron. Agric. 121, 313–319.

Mathew, S., Snigdha, S., Mathew, J., Radhakrishnan, E.K., 2019. Biodegradable and active nanocomposite pouches reinforced with silver nanoparticles for improved packaging of chicken sausages. Food Packag. Shelf Life 19, 155–166.

Medeiros, E.S., Gregório, R., Martinez, R.A., Mattoso, L.H.C., 2009. A taste sensor array based on polyaniline nanofibers for orange juice quality assessment. Sens. Lett. 7, 24–30.

Meriç, S., Çakır, Ö., Turgut-Kara, N., Arı, S., 2014. Detection of genetically modified maize and soybean in feed samples. Genet. Mol. Res. 13, 1160–1168.

Miranda-Castro, R., de los Santos Álvarez, N., Lobo-Castañón, M.J., 2020. Aptamer-based magnetoassay for gluten determination. In: Laboratory Methods in Dynamic Electroanalysis. Elsevier, Amsterdam, the Netherlands, pp. 221–229.

Mishra, G.K., Barfidokht, A., Tehrani, F., Mishra, R.K., 2018. Food safety analysis using electrochemical biosensors. Foods 7, 141.

Mishra, R.K., Hayat, A., Catanante, G., Ocaña, C., Marty, J.-L., 2015. A label free aptasensor for ochratoxin a detection in cocoa beans: an application to chocolate industries. Anal. Chim. Acta 889, 106–112.

Mitura, K., Wyrebski, K., Zarzycki, P.K., 2017. Bioactive food packaging with nanodiamond particles manufactured by detonation and plasma-chemical methods. In: Food Packaging. Elsevier, Amsterdam, The Netherlands, pp. 295–328. ISBN 978-0-12-804302-8.

Mohammad, R., Ahmad, M., Heng, L.Y., 2014. Chilli hotness determination based on optical capsaicin biosensor using stacked immobilisation technique. Sensors Actuators B Chem. 190, 593–600.

Mohammad-Razdari, A., Ghasemi-Varnamkhasti, M., Rostami, S., Izadi, Z., Ensafi, A.A., Siadat, M., 2020. Development of an electrochemical biosensor for impedimetric detection of tetracycline in milk. J. Food Sci. Technol. 57, 4697–4706.

Moran, K.L.M., Fitzgerald, J., McPartlin, D.A., Loftus, J.H., O'Kennedy, R., 2016. Biosensor-based technologies for the detection of pathogens and toxins. In: Comprehensive Analytical Chemistry. vol. 74. Elsevier, Amsterdam, The Netherlands, pp. 93–120. ISBN 978-0-444-63579-2.

Mourot, B.P., Gruffat, D., Durand, D., Chesneau, G., Mairesse, G., Andueza, D., 2015. Breeds and muscle types modulate performance of near-infrared reflectance spectroscopy to predict the fatty acid composition of bovine meat. Meat Sci. 99, 104–112.

Mustafa, F., Andreescu, S., 2018. Chemical and biological sensors for food-quality monitoring and smart packaging. Foods 7, 168.

Mustafa, F., Hassan, R.Y.A., Andreescu, S., 2017. Multifunctional nanotechnology-enabled sensors for rapid capture and detection of pathogens. Sensors 17, 2121.

Neethirajan, S., Freund, M.S., Jayas, D.J., 2013. Poly(Aniline Boronic Acid) Polymers and Methods of Use. United States Patent 8454819 B2.

Neethirajan, S., Freund, M., Jayas, D., Shafai, C., Thomson, D., White, N., 2010. Development of carbon dioxide (CO2) sensor for grain quality monitoring. Biosyst. Eng. 106, 395–404.

Neethirajan, S., Ragavan, V., Weng, X., Chand, R., 2018a. Biosensors for sustainable food engineering: challenges and perspectives. Biosensors 8, 23.

Neethirajan, S., Weng, X., Tah, A., Cordero, J.O., Ragavan, K.V., 2018b. Nano-biosensor platforms for detecting food allergens—new trends. Sens. Bio Sens. Res. 18, 13–30.

Németh, D., Balazs, G., Daood, H.G., Kovacs, Z., Bodor, Z., Zaukuu, J.L.Z., Szentpeteri, V., Kokai, Z., Kappel, N., 2019. Standard analytical methods, sensory evaluation, NIRS and electronic tongue for sensing taste attributes of different melon varieties. Sensors 19, 5010.

Neufeld, T., Schwartz-Mittelmann, A., Biran, D., Ron, E.Z., Rishpon, J., 2003. Combined phage typing and amperometric detection of released enzymatic activity for the specific identification and quantification of bacteria. Anal. Chem. 75, 580–585.

Nile, S.H., Baskar, V., Selvaraj, D., Nile, A., Xiao, J., Kai, G., 2020. Nanotechnologies in food science: applications, recent trends, and future perspectives. Nano Micro Lett. 12, 45.

Olsen, E.V., Sorokulova, I.B., Petrenko, V.A., Chen, I.H., Barbaree, J.M., Vodyanoy, V.J., 2006. Affinity-selected filamentous bacteriophage as a probe for acoustic wave biodetectors of Salmonella typhimurium. Biosens. Bioelectron. 21, 1434–1442.

Oroian, M., Paduret, S., Ropciuc, S., 2018. Honey adulteration detection: voltammetric e-tongue versus official methods for physicochemical parameter determination. J. Sci. Food Agric. 98, 4304–4311.

Pandey, V.K., Upadhyay, S.N., Niranjan, K., Mishra, P.K., 2020. Antimicrobial biodegradable chitosan-based composite nano-layers for food packaging. Int. J. Biol. Macromol. 157, 212–219.

Pastucha, M., Farka, Z., Lacina, K., Mikušová, Z., Skládal, P., 2019. Magnetic nanoparticles for smart electrochemical immunoassays: a review on recent developments. Microchim. Acta 186, 312.

Patra, S., Roy, E., Madhuri, R., Sharma, P.K., 2017. A technique comes to life for security of life: the food contaminant sensors. In: Nanobiosensors. Elsevier, Amsterdam, The Netherlands, pp. 713–772. ISBN 978-0-12-804301-1.

Petrovic, G., Aleixandre-Tudo, J.L., Buica, A., 2020. Viability of IR spectroscopy for the accurate measurement of yeast assimilable nitrogen content of grape juice. Talanta 206, 120241.

Pinto, G., Ferranti, P., 2016. Impact of climate change on the food chain. In: Reference Module in Food Science. Elsevier, Amsterdam, The Netherlands. ISBN 978-0-08-100596-5.

Prieto, N., López-campos, Ó., Aalhus, J.L., Dugan, M.E.R., Juárez, M., Uttaro, B., 2014. Use of near infrared spectroscopy for estimating meat chemical composition, quality traits and fatty acid content from cattle fed sun flower or flaxseed. Meat Sci. 98, 279–288.

Primožič, M., Knez, Z., Leitgeb, M., 2021. (Bio)nanotechnology in food science—food packaging. Nanomaterials 11, 292.

Que, X., Chen, X., Fu, L., Lai, W., Zhuang, J., Chen, G., Tang, D., 2013a. Platinum-catalyzed hydrogen evolution reaction for sensitive electrochemical immunoassay of tetracycline residues. J. Electroanal. Chem. 704, 111–117.

Que, X., Liu, B., Fu, L., Zhuang, J., Chen, G., Tang, D., 2013b. Molecular imprint for electrochemical detection of streptomycin residues using enzyme signal amplification. Electroanalysis 25, 531–537.

Rady, A., Adedeji, A., 2018. Assessing different processed meats for adulterants using visible-near-infrared spectroscopy. Meat Sci. 136, 59–67.

Raja, I.S., Vedhanayagam, M., Preeth, D.R., Kim, C., Lee, J.H., Han, D.W., 2021. Development of two-dimensional nanomaterials based electrochemical biosensors on enhancing the analysis of food toxicants. Int. J. Mol. Sci. 22 (6), 3277.

Ranjan, S., Dasgupta, N., Chakraborty, A., Melvin Samuel, S., Ramalingam, C., Shanker, R., Kumar, A., 2014. Nanoscience and nanotechnologies in food industries: opportunities and research trends. J. Nanopart. Res. 16, 1–23.

Rapini, R., Marrazza, G., 2017. Electrochemical aptasensors for contaminants detection in food and environment: recent advances. Bioelectrochemistry 118, 47–61.

Różańska, A., Dymerski, T., Namieśnik, J., 2018. Novel analytical method for detection of orange juice adulteration based on ultra-fast gas chromatography. Chem. Mon. 149, 1615–1621.

Sa, N.M.S.M., Mattos, A.L.A., Silva, L.M.A., Brito, E.S., Rosa, M.F., Azeredo, H.M.C., 2020. From cashew byproducts to biodegradable active materials: bacterial cellulose-lignin-cellulose nanocrystal nanocomposite films. Int. J. Biol. Macromol. 161, 1337–1345.

Santos, P.V.F., Lopes, I.C., Diculescu, V.C., Oliveira-Brett, A.M., 2012. DNA—cyanobacterial hepatotoxins microcystin-LR and nodularin interaction: electrochemical evaluation. Electroanalysis 24, 547–553.

Santos, A.S., Pereira, A.C., Durán, N., Kubota, L.T., 2006. Amperometric biosensor for ethanol based on co-immobilization of alcohol dehydrogenase and meldola's blue on multi-wall carbon nanotube. Electrochim. Acta 52, 215–220.

Saucedo, N.M., Mulchandani, A., 2016. Sensing of biological contaminants. In: Comprehensive Analytical Chemistry. vol. 74. Elsevier, Amsterdam, The Netherlands, pp. 73–91. ISBN 978-0-444-63579-2.

Scarano, S., Mariani, S., Minunni, M., 2016. Label free affinity sensing: application to food analysis. Acta IMEKO 5, 36–44.

Scognamiglio, V., Antonacci, A., Lambreva, M.D., Litescu, S.C., Rea, G., 2015. Synthetic biology and biomimetic chemistry as converging technologies fostering a new generation of smart biosensors. Biosens. Bioelectron. 74, 1076–1086.

Selva Kumar, L.S., Thakur, M.S., 2011. Competitive immunoassay for analysis of vitamin B(12). Anal. Biochem. 418, 238–246.

Setterington, E.B., Alocilja, E.C., 2012. Electrochemical biosensor for rapid and sensitive detection of magnetically extracted bacterial pathogens. Biosensors 2, 15.

Shabani, A., Marquette, C.A., Mandeville, R., Lawrence, M.F., 2013. Magnetically-assisted impedimetric detection of bacteria using phage-modified carbon microarrays. Talanta 116, 1047–1053.

Sharma, A., Istamboulie, G., Hayat, A., Catanante, G., Bhand, S., Marty, J.L., 2017. Disposable and portable aptamer functionalized impedimetric sensor for detection of kanamycin residue in milk sample. Sensors Actuators B Chem. 245, 507–515.

Shevade, A.V., Ryan, M.A., Homer, M.L., Manfreda, A.M., Zhou, H., Manatt, K.S., 2003. Molecular modeling of polymer composite–analyte interactions in electronic nose sensors. Sensors Actuators B Chem. 93, 84–91.

Silva, N.F., Magalhães, J.M., Barroso, M.F., Oliva-Teles, T., Freire, C., Delerue-Matos, C., 2019. In situ formation of gold nanoparticles in polymer inclusion membrane: application as platform in a label-free potentiometric immunosensor for Salmonella typhimurium detection. Talanta 194, 134–142.

Silva, N.F.D., Magalhães, J.M.C.S., Freire, C., Delerue-Matos, C., 2018. Electrochemical biosensors for Salmonella: state of the art and challenges in food safety assessment. Biosens. Bioelectron. 99, 667–682.

Siva balan, K.C., 2016. Robotic-based agriculture for rural renaissance. In: Comprehensive Analytical Chemistry. vol. 74. Elsevier, Amsterdam, The Netherlands, pp. 363–375. ISBN 978-0-444-63579-2.

Soares, R.R.A., Hjort, R.G., Pola, C.C., Parate, K., Reis, E.L., Soares, N.F.F., McLamore, E.S., Claussen, J.C., Gomes, C.L., 2020. Laser-induced graphene electrochemical immunosensors for rapid and label-free monitoring of salmonella enterica in chicken broth. ACS Sens. 5, 1900–1911.

Sobrino-Gregorio, L., Tanleque-Alberto, F., Bataller, R., Soto, J., Escriche, I., 2020. Using an automatic pulse voltammetric electronic tongue to verify the origin of honey from Spain, Honduras, and Mozambique. J. Sci. Food Agric. 100, 212–217.

Song, H.S., Kwon, O.S., Lee, S.H., Park, S.J., Kim, U.K., Jang, J., Park, T.H., 2012. Human taste receptor-functionalized field effect transistor as a human-like nanobioelectronic tongue. Nano Lett. 13, 172–178.

Soto-Barajas, M.C., González-Martín, M.I., Salvador-Esteban, J., Hernández-Hierro, J.M., Moreno-Rodilla, V., Vivar-Quintana, A.M., Revilla, I., Ortega, I.L., Morón-Sancho, R., Curto-Diego, B., 2013. Prediction of the type of milk and degree of ripening in cheeses by means of artificial neural networks with data concerning fatty acids and near infrared spectroscopy. Talanta 116, 50–55.

Spano, G., Russo, P., Lonvaud-Funel, A., Lucas, P., Alexandre, H., Grandvalet, C., Coton, E., Coton, M., Barnavon, L., Bach, B., et al., 2010. Biogenic amines in fermented foods. Eur. J. Clin. Nutr. 64, S95–S100.

Stadler, R.H., 2016. Foreword for food processing—derived contaminants in food analysis. In: Reference Module in Food Science. Elsevier, Amsterdam, The Netherlands. ISBN 978-0-08-100596-5.

Stadler, R.H., Blank, I., Varga, N., Robert, F., Hau, J., Guy, P.A., Robert, M.-C., Riediker, S., 2002. Food chemistry: acrylamide from Maillard reaction products. Nature 419, 449–450.

Sun, Y., Wang, Y., Huang, J., Ren, G., Ning, J., Deng, W., Li, L., Zhang, Z., 2020. Quality assessment of instant green tea using portable NIR spectrometer. Spectrochim. Acta A Mol. Biomol. Spectrosc. 240, 118576.

Sundramoorthy, A.K., Gunasekaran, S., 2014. Applications of graphene in quality assurance and safety of food. TrAC Trends Anal. Chem. 60, 36–53.

Taghdisi, S.M., Danesh, N.M., Ramezani, M., Abnous, K., 2016. A novel m-shape electrochemical aptasensor for ultrasensitive detection of tetracyclines. Biosens. Bioelectron. 85, 509–514.

Tahir, H.E., Arslan, M., Mahunu, G.K., Jiyong, S., Xiaobo, Z., Gasmalla, M.A.A., Mariod, A.A., 2019. Data fusion approach improves the prediction of single phenolic compounds in honey: a study of NIR and Raman spectroscopies. eFood 1, 173–180.

Tanner, D., 2012. Applications for RFID technologies in the food supply chain. In: Reference Module in Food Science. Elsevier, Amsterdam, The Netherlands. ISBN 978-0-08-100596-5.

Tawil, N., Sacher, E., Mandeville, R., Meunier, M., 2012. Surface plasmon resonance detection of *E. coli* and methicillin-resistant *S. aureus* using bacteriophages. Biosens. Bioelectron. 37, 24–29.

Thavarungkul, P., Dawan, S., Kanatharana, P., Asawatreratanakul, P., 2007. Detecting penicillin G in milk with impedimetric label-free immunosensor. Biosens. Bioelectron. 23, 688–694.

Tian, X., Wang, J., Cui, S., 2013. Analysis of pork adulteration in minced mutton using electronic nose of metal oxide sensors. J. Food Eng. 119, 744–749.

Titoiu, A.M., Porumb, R., Fanjul-Bolado, P., Epure, P., Zamfir, M., Vasilescu, A., 2019. Detection of allergenic lysozyme during winemaking with an electrochemical aptasensor. Electroanalysis 31, 2262–2273.

Tokonami, S., Nakadoi, Y., Takahashi, M., Ikemizu, M., Kadoma, T., Saimatsu, K., Dung, L.Q., Shiigi, H., Nagaoka, T., 2013. Label-free and selective bacteria detection using a film with transferred bacterial configuration. Anal. Chem. 85, 4925–4929.

Trihaas, J., Vognsen, L., Nielsen, P.V., 2005. Electronic nose: new tool in modelling the ripening of Danish blue cheese. Int. Dairy J. 15, 679–691.

Triyana, K., Taukhid Subekti, M., Aji, P., Nur Hidayat, S., Rohman, A., 2015. Development of electronic nose with low-cost dynamic headspace for classifying vegetable oils and animal fats. Appl. Mech. Mater. 771, 50–54.

Tyagi, D., Wang, H., Huang, W., Hu, L., Tang, Y., Guo, Z., Ouyang, Z., Zhang, H., 2020. Recent advances in two-dimensional material-based sensing technology toward health and environmental monitoring applications. Nanoscale 12, 3535–3559.

U.S. Food and Drug Administration (USFDA), 2011. Fish and Fishery Products Hazards and Controls Guidance, fourth ed. Center for Food Safety and Applied Nutrition, College Park, MD, USA.

Valat, C., Limoges, B., Huet, D., Romette, J.-L., 2000. A disposable protein A-based immunosensor for flow-injection assay with electrochemical detection. Anal. Chim. Acta 404, 187–194.

Valdes, A., Mellinas, A.C., Ramos, M., Burgos, N., Jimenez, A., Garrigos, M.C., 2015. Use of herbs, spices and their bioactive compounds in active food packaging. RSC Adv. 5, 40324–40335.

Vanderroost, M., Ragaert, P., Devlieghere, F., De Meulenaer, B., 2014. Intelligent food packaging: the next generation. Trends Food Sci. Technol. 39, 47–62.

Vanegas, D.C., Gomes, C.L., Cavallaro, N.D., Giraldo-Escobar, D., McLamore, E.S., 2017. Emerging biorecognition and transduction schemes for rapid detection of pathogenic bacteria in food. Compr. Rev. Food Sci. Food Saf. 16, 1188–1205.

Vanegas, D.C., Patiño, L., Mendez, C., de Oliveira, D.A., Torres, A.M., Gomes, C.L., McLamore, E.S., 2018. Laser scribed graphene biosensor for detection of biogenic amines in food samples using locally sourced materials. Biosensors 8, 42.

Viljoen, M., Hoffman, L.C., Brand, T.S., 2005. Prediction of the chemical composition of freeze dried ostrich meat with near infrared reflectance spectroscopy. Meat Sci. 69, 255–261.

Visciano, P., Schirone, M., Tofalo, R., Suzzi, G., 2014. Histamine poisoning and control measures in fish and fishery products. Front. Microbiol. 5, 500.

Visconti, L.G., Rodríguez, M.S., Di Anibal, C.V., 2020. Determination of grated hard cheeses adulteration by near infrared spectroscopy (NIR) and multivariate analysis. Int. Dairy J. 104, 104647.

Vogiazi, V., de la Cruz, A., Mishra, S., Shanov, V., Heineman, W.R., Dionysiou, D.D., 2019. A comprehensive review: development of electrochemical biosensors for detection of cyanotoxins in freshwater. ACS Sens. 4, 1151–1173.

Wang, C., Chang, T., Dong, S., Zhang, D., Ma, C., Chen, S., Li, H., 2020. Biopolymer films based on chitosan/potato protein/linseed oil/ZnO NPs to maintain the storage quality of raw meat. Food Chem. 332, 127375.

Wang, P., Lin, Z., Su, X., Tang, Z., 2017. Application of au based nanomaterials in analytical science. Nano Today 12, 64–97.

Wang, J., Zareef, M., He, P., Sun, H., Chen, Q., Li, H., Ouyang, Q., Guo, Z., Zhang, Z., Xu, D., 2019. Evaluation of matcha tea quality index using portable NIR spectroscopy coupled with chemometric algorithms. J. Sci. Food Agric. 99, 5019–5027.

Wang, H., Zhou, Y., Jiang, X., Sun, B., Zhu, Y., Wang, H., Su, Y., He, Y., 2015. Simultaneous capture, cetection, and inactivation of bacteria as enabled by a surface-enhanced raman scattering multifunctional chip. Angew. Chem. Int. Ed. 54, 5132–5136.

Wei, Z., Wang, J., 2011. Detection of antibiotic residues in bovine milk by a voltammetric electronic tongue system. Anal. Chim. Acta 694, 46–56.

Wei, Z., Zhang, W., Wang, Y., Wang, J., 2017. Monitoring the fermentation, post-ripeness and storage processes of set yogurt using voltammetric electronic tongue. J. Food Eng. 203, 41–52.

Wei, Q., Zhao, Y., Du, B., Wu, D., Li, H., Yang, M., 2012. Ultrasensitive detection of kanamycin in animal derived foods by label-free electrochemical immunosensor. Food Chem. 134, 1601–1606.

Weng, X., Neethirajan, S., 2016. A micro fluidic biosensor using graphene oxide and aptamer-functionalized quantum dots for peanut allergen detection. Biosens. Bioelectron. 85, 649–656. Biosensors and Bioelectronics.

Weng, X., Neethirajan, S., 2017. Ensuring food safety: quality monitoring using microfluidics. Trends Food Sci. Technol. 65, 10–22.

White, S.P., Frisbie, C.D., Dorfman, K.D., 2018. Detection and sourcing of gluten in grain with multiple floating-gate transistor biosensors. ACS Sens. 3, 395–402.

Wong, D.E., Dai, M., Talbert, J.N., Nugen, S.R., Goddard, J.M., 2014. Biocatalytic polymer nanofibers for stabilization and delivery of enzymes. J. Mol. Catal. B 110, 16–22.

World Health Organization (WHO), 2015. WHO Estimates of the Global Burden of Foodborne Diseases: Foodborne Diseases Burden Epidemiology Reference Group 2007–2015. WHO Report, WHO, Geneva, Switzerland, p. 255.

Wu, H., Aoki, A., Arimoto, T., Nakano, T., Ohnuki, H., Murata, M., Ren, H., Endo, H., 2015. Fish stress become visible: a new attempt to use biosensor for real-time monitoring fish stress. Biosens. Bioelectron. 67, 503–510.

Wu, S., Duan, N., Shi, Z., Fang, C., Wang, Z., 2014. Simultaneous aptasensor for multiplex pathogenic bacteria detection based on multicolor upconversion nanoparticles labels. Anal. Chem. 86, 3100–3107.

Wu, Z., Huang, Y., Xiao, L., Lin, D., Yang, Y., Wang, H., Yang, Y., Wu, D., Chen, H., Zhang, Q., et al., 2019a. Physical properties and structural characterization of starch/polyvinyl alcohol/graphene oxide composite films. Int. J. Biol. Macromol. 123, 569–575.

Wu, Y., Luo, Y., Wang, Q., 2012. Antioxidant and antimicrobial properties of essential oils encapsulated in zein nanoparticles prepared by liquid–liquid dispersion method. Food Sci. Technol. 48, 283–290.

Wu, Q., Zhang, Y., Yang, Q., Yuan, N., Zhang, W., 2019b. Review of electrochemical DNA biosensors for detecting food borne pathogens. Sensors (Basel) 19 (22), 4916.

Xia, F., Wang, H., Xiao, D., Dubey, M., Ramasubramaniam, A., 2014. Two-dimensional material nanophotonics. Nat. Photonics 8, 899–907.

Xu, M., Wang, J., Zhu, L., 2019. The qualitative and quantitative assessment of tea quality based on E-nose, E-tongue and E-eye combined with chemometrics. Food Chem. 289, 482–489.

Xu, L., Yu, X., Liu, L., Zhang, R., 2016. A novel method for qualitative analysis of edible oil oxidation using an electronic nose. Food Chem. 202, 229–235.

Yan, Z., Gan, N., Li, T., Cao, Y., Chen, Y., 2016. A sensitive electrochemical aptasensor for multiplex antibiotics detection based on high-capacity magnetic hollow porous nanotracers coupling exonuclease-assisted cascade target recycling. Biosens. Bioelectron. 78, 51–57.

Yan, L., Luo, C., Cheng, W., Mao, W., Zhang, D., Ding, S., 2012. A simple and sensitive electrochemical aptasensor for determination of chloramphenicol in honey based on target-induced strand release. J. Electroanal. Chem. 687, 89–94.

Yan, Z., Niu, Q., Mou, M., Wu, Y., Liu, X., Liao, S., 2017. A novel colorimetric method based on copper nanoclusters with intrinsic peroxidase-like for detecting xanthine in serum samples. J. Nanopart. Res. 19, 235.

Yang, G., Zhao, F., 2015. Electrochemical sensor for chloramphenicol based on novel multiwalled carbon nanotubes@molecularly imprinted polymer. Biosens. Bioelectron. 64, 416–422.

Yang, T., Huang, H., Zhu, F., Lin, Q., Zhang, L., Liu, J., 2016. Recent progresses in nanobiosensing for food safety analysis. Sensors 16, 1118.

Yang, J.Y., Li, Y., Chen, S.M., Lin, K.C., 2011. Fabrication of a cholesterol biosensor based on cholesterol oxidase and multiwall carbon nanotube hybrid composites. Int. J. Electrochem. Sci. 6, 2223–2224.

Yang, Z., Wang, Y., Zhang, D., 2017. A novel multifunctional electrochemical platform for simultaneous detection, elimination, and inactivation of pathogenic bacteria based on the vancomycin-functionalised AgNPs/3d-ZnO nanorod arrays. Biosens. Bioelectron. 98, 248–253.

Yang, R., Zhao, J., Chen, M., Yang, T., Luo, S., Jiao, K., 2015. Electrocatalytic determination of chloramphenicol based on molybdenum disulfide nanosheets and self-doped polyaniline. Talanta 131, 619–623.

Yilmaz, A.H., 2020. Antibacterial activity of chitosan-based systems. Funct. Chitosan, 457–489.

Yilmaz, E., Majidi, D., Ozgur, E., Denizli, A., 2015. Whole cell imprinting based *Escherichia coli* sensors: a study for SPR and QCM. Sensors Actuators B Chem. 209, 714–721.

Yin, J., Guo, W., Qin, X., Pei, M., Wang, L., Ding, F., 2016a. A regular "signal attenuation" electrochemical aptasensor for highly sensitive detection of streptomycin. New J. Chem. 40, 9711–9718.

Yin, Y., Qin, X., Wang, Q., Yin, Y., 2016b. A novel electrochemical aptasensor for sensitive detection of streptomycin based on gold nanoparticle-functionalized magnetic multi-walled carbon nanotubes and nanoporous ptti alloy. RSC Adv. 6, 39401–39408.

Yola, M.L., Uzun, L., Ozaltin, N., Denizli, A., 2014. Development of molecular imprinted nanosensor for determination of tobramycin in pharmaceuticals and foods. Talanta 120, 318–324.

Yu, X., Chen, F., Wang, R., Li, Y., 2018. Whole-bacterium SELEX of DNA aptamers for rapid detection of *E. coli* O157: H7 using a QCM sensor. J. Biotechnol. 266, 39–49.

Yu, S., Wei, Q., Du, B., Wu, D., Li, H., Yan, L., Ma, H., Zhang, Y., 2013. Label-free immunosensor for the detection of kanamycin using Ag@Fe3O4 nanoparticles and thionine mixed graphene sheet. Biosens. Bioelectron. 48, 224–229.

Zambrano-Zaragoza, M.L., González-Reza, R., Mendoza-Muñoz, N., Miranda-Linares, V., Bernal-Couoh, T.F., Mendoza-Elvira, S., Quintanar-Guerrero, D., 2018. Nanosystems in edible coatings: a novel strategy for food preservation. Int. J. Mol. Sci. 19, 705.

Zamora-Rojas, E., Garrido-Varo, A., de Pedro-Sanz, E., Guerrero-Ginel, J.E., Pérez-Marín, D., 2013. Prediction of fatty acids content in pig adipose tissue by near infrared spectroscopy: at-line versus in-situ analysis. Meat Sci. 95, 503–511.

Zelada-Guillén, G.A., Riu, J., Düzgün, A., Rius, F.X., 2009. Immediate detection of living bacteria at ultralow concentrations using a carbon nanotube based potentiometric aptasensor. Angew. Chem. Int. Ed. 48, 7334–7337.

Zhan, X., Hu, G., Wagberg, T., Zhan, S., Xu, H., Zhou, P., 2016. Electrochemical aptasensor for tetracycline using a screen-printed carbon electrode modified with an alginate film containing reduced graphene oxide and magnetite (Fe3O4) nanoparticles. Microchim. Acta 183, 723–729.

Zhang, Y., Chen, X., 2019. Nanotechnology and nanomaterials-based no-wash electrochemical biosensors: from design to application. Nanoscale 11, 19105–19118.

Zhang, Z., Tong, J., Chen, D., Lan, Y., 2008. Electronic nose with an air sensor matrix for detecting beef freshness. J. Bionic Eng. 5, 67–73.

Zhang, N., Xiao, F., Bai, J., Lai, Y., Hou, J., Xian, Y., Jin, L., 2011. Label-free immunoassay for chloramphenicol based on hollow gold nanospheres/chitosan composite. Talanta 87, 100–105.

Zhang, X., Zhang, Y.-C., Zhang, J.-W., 2016. A highly selective electrochemical sensor for chloramphenicol based on three-dimensional reduced graphene oxide architectures. Talanta 161, 567–573.

Zhou, L., Li, D.-J., Gai, L., Wang, J.-P., Li, Y.-B., 2012. Electrochemical aptasensor for the detection of tetracycline with multi-walled carbon nanotubes amplification. Sensors Actuators B Chem. 162, 201–208.

Zhu, Y., Chandra, P., Song, K.-M., Ban, C., Shim, Y.-B., 2012. Label-free detection of kanamycin based on the aptamer-functionalized conducting polymer/gold nanocomposite. Biosens. Bioelectron. 36, 29–34.

Zhu, X., Li, S., Shan, Y., Zhang, Z., Li, G., Su, D., Liu, F., 2010. Detection of adulterants such as sweeteners materials in honey using near-infrared spectroscopy and chemometrics. J. Food Eng. 101, 92–97.

Index

Note: Page numbers followed by *f* indicate figures.

CPI Antony Rowe
Eastbourne, UK
November 22, 2023